AF863921

Reports of China's Basic Research

Editor-in-Chief

Wei Yang, National Natural Science Foundation of China, Beijing, China, Zhejiang University, Hangzhou, Zhejiang, China

The National Natural Science Foundation of China (NSFC) was established on February 14, 1986. Upon its establishment, NSFC was an institution directly under the jurisdiction of the State Council, tasked with the administration of the National Natural Science Fund from the Central Government. In 2018, it became managed by the Ministry of Science and Technology (MOST) but kept its due independence in operation. Since its establishment, NSFC has comprehensively introduced and implemented a rigorous and objective merit-review system to fulfill its mission of supporting basic research, fostering talented researchers, developing international cooperation and promoting socioeconomic development.

Featuring science, basics, and advances, the series of *Reports of China's Basic Research* is organized by the NSFC to present the overall level and pattern of China's basic research, share innovative achievements, and illustrate excellent breakthroughs in key fields. It covers various disciplines including but not limited to, computer science, materials science, life sciences, engineering, environmental sciences, mathematics, and physics. The series will show the core contents of the final reports of the Major Programs and the Major Research Plans funded by NSFC, and will closely follow the frontiers of basic research developments in China.

If you are interested in publishing your book in the series, please contact Qian Xu (Email: violetta_xuqian@zju.edu.cn) and Mengchu Huang (Email: mengchu.huang@cn.springer.com).

Maochun Hong
Editor

Structural Design and Controllable Preparation of the Function-Directed Crystalline Materials

ZHEJIANG UNIVERSITY PRESS
浙江大学出版社

Springer

Editor
Maochun Hong
Fujian Institute of Research
on the Structure of Matter
Chinese Academy of Sciences
Fuzhou, China

ISSN 2731-8907 ISSN 2731-8915 (electronic)
Reports of China's Basic Research
ISBN 978-981-99-3767-7 ISBN 978-981-99-3768-4 (eBook)
https://doi.org/10.1007/978-981-99-3768-4

Jointly published with Zhejiang University Press
The print edition is not for sale in China (Mainland). Customers from China (Mainland) please order the print book from: Zhejiang University Press.

© Zhejiang University Press 2023

This work is subject to copyright. All rights are solely and exclusively licensed by the Publisher, whether the whole or part of the material is concerned, specifically the rights of reprinting, reuse of illustrations, recitation, broadcasting, reproduction on microfilms or in any other physical way, and transmission or information storage and retrieval, electronic adaptation, computer software, or by similar or dissimilar methodology now known or hereafter developed.
The use of general descriptive names, registered names, trademarks, service marks, etc. in this publication does not imply, even in the absence of a specific statement, that such names are exempt from the relevant protective laws and regulations and therefore free for general use.
The publishers, the authors, and the editors are safe to assume that the advice and information in this book are believed to be true and accurate at the date of publication. Neither the publishers nor the authors or the editors give a warranty, expressed or implied, with respect to the material contained herein or for any errors or omissions that may have been made. The publishers remain neutral with regard to jurisdictional claims in published maps and institutional affiliations.

This Springer imprint is published by the registered company Springer Nature Singapore Pte Ltd.
The registered company address is: 152 Beach Road, #21-01/04 Gateway East, Singapore 189721, Singapore

Paper in this product is recyclable.

Editorial Board

Editor-in-Chief

Wei Yang

Associate Editors

Ruiping Gao
Yu Han

Editors

Sheng Yu
Changrui Wang
Qidong Wang
Shumei Lü
Weitong Zhu
Ke Liu
Zuoyi Liu
Ruijuan Sun
Junlin Yang
Liyao Zou
Xiangping Zhang
Yongjun Chen
Yupeng Yao
Yanying Xu
Jianquan Guo

Jie Peng
Guoxuan Dong
Zhiyong Han

Preface to the Series

As Lao Tzu said, "A huge tree grows from a tiny seedling; a nine-storied tower rises from a heap of earth." Basic research is the fundamental approach to fostering innovation-driven development, and its level becomes an important yardstick for measuring the overall scientific and national strength of a country. Since the beginning of the twenty-first century, China's overall strength in basic research has been consistently increasing. With respect to input and output, China's input in basic research increased by 14.8 times from 5.22 billion yuan in 2001 to 82.29 billion yuan in 2016, with an average annual increase of 20.2%. In the same period, the number of China's scientific papers included in the Science Citation Index (SCI) increased from lower than 40,000 to 324,000; China rose from the 6th to the 2nd place in global ranking in terms of the number of published papers. In regard to the quality of output, in 2016, China ranked No. 2 in the world in terms of citations in 9 disciplines, among which materials science ranked No. 1; as of October 2017, China ranked No. 3 in the world in the number of both Highly Cited Papers (top 1%) and Hot Papers (top 0.1%), with the latter accounting for 25.1% of the global total. In talent cultivation, in 2006, China had 175 scientists (136 of whom from the Chinese mainland) included in Thomson Reuters' list of Highly Cited Researchers, ranking 4th globally and 1st in Asia.

Meanwhile, we should also be keenly aware that China's basic research is still facing great challenges. First, funding for basic research in China is still far less than that in developed countries—only about 5% of the R&D funds in China are used for basic research, a much lower percentage than 15%–20% in developed countries. Second, competence for original innovation in China is insufficient. Major original scientific achievements that have global impact are still rare. Most of the scientific research projects are just a follow-up or imitation of existing research, rather than groundbreaking research. Third, the development of disciplines is not balanced, and China's research level in some disciplines is noticeably lower than the international level—China's Field-Weighted Citation Impact (FWCI) in disciplines just reached 0.94 in 2016, lower than the world average of 1.0.

The Chinese government attaches great importance to basic research. In the 13th Five-Year Plan (2016–2020), China has established scientific and technological innovation as a priority in all-round innovation and has made strategic arrangements to strengthen basic research. General Secretary XI Jinping put forward a grand blueprint of making China into a world-leading power in science and technology in his speech delivered at the National Conference on Scientific and Technological Innovation in 2016, and emphasized that "we should aim for the frontiers of science and technology, strengthen basic research, and make major breakthroughs in pioneering basic research and groundbreaking and original innovations" at the 19th CPC National Congress on Oct. 18, 2017. With more than 30 years of unremitting exploration, the National Natural Science Foundation of China (NSFC), one of the main channels for supporting basic research in China, has gradually shaped a funding pattern covering research, talent, tools, and convergence, and has taken action to vigorously promote basic frontier research and the growth of scientific research talent, reinforce the building of innovative research teams, deepen regional cooperation and exchanges, and push forward multidisciplinary convergence. As of 2016, nearly 70% of China's published scientific papers were funded by the NSFC, accounting for 1/9 of the total number of published papers all over the world. Facing the new strategic target of building China into a strong country in science and technology, the NSFC will conscientiously reinforce forward-looking planning and enhance the efficiency of evaluation, so as to achieve the strategic goal of making China progressively share the same level with major innovative countries in research total volume, contribution, and groundbreaking researchers by 2050.

The series *Advances in China's Basic Research* and the series *Reports of China's Basic Research* proposed and planned by the NSFC emerge against such a background. Featuring science, basics, and advances, the two series are aimed at sharing innovative achievements, diffusing performances of basic research, and leading breakthroughs in key fields. They closely follow the frontiers of basic research developments in China and publish excellent innovation achievements funded by the NSFC. The series of *Advances in China's Basic Research* mainly presents the important original achievements of the programs funded by the NSFC and demonstrates the breakthroughs and forward guidance in key research fields; the series *Reports of China's Basic Research* shows the core contents of the final reports of Major Programs and Major Research Plans funded by the NSFC to make a systematic summarization and give a strategic outlook on the achievements in the funding priorities of the NSFC. We hope not only to comprehensively and systematically introduce backgrounds, scientific significance, discipline layouts, frontier breakthroughs of the programs, and a strategic outlook for the subsequent research, but also to summarize innovative ideas, enhance multidisciplinary convergence, foster the continuous development of research in concerned fields, and promote original discoveries.

As Hsun Tzu remarked, "When earth piles up into a mountain, wind and rain will originate thereof. When waters accumulate into a deep pool, dragons will come to live in it." The series *Advances in China's Basic Research* and *Reports of China's Basic Research* are expected to become the "historical records" of China's basic research. They will provide researchers with abundant scientific research material and vitality of innovation, and will certainly play an active role in making China's basic research prosper and building China's strength in science and technology.

Wei Yang
Academician of the Chinese
Academy of Sciences
Beijing, China

Preface

As the foundation and precursor of modern science and technology, materials science is an internationally recognized core field. Materials science is intertwined with national economy, engineering and technology, and national security, which is critical to the national social and economic development. Developed countries have prioritized novel materials as a key area in the development of science and technology to maintain their economic and technological leadership.

The invention and application of new materials are milestones of human civilization, and the development of materials science has contributed to the advancement of human society and civilization. People have long expected a new paradigm of materials engineering to guide the search for new materials, i.e., to predict material properties based on known scientific laws or to design and prepare new materials with specific functions according to the properties required, thereby shortening new materials' development cycles. So far, to develop new materials scientists have mastered the correlations between macroscopic properties and structures (molecular and spatial structures) of materials, further explored the intrinsic laws, and systematically discussed such relationship based on quantum and structural chemistry.

Materials' functional properties are primarily derived from their optical, electrical, magnetic, thermal, and mechanical effects or combinations of these effects (e.g., photovoltaic, electro-optical, acoustic-optical, magnetoelectric, and thermoelectric effects); and their research and development have shifted the focus from a single-functional to multifunctional one. It is well-known that a material's Photoelectromagnetic (PEM) properties are mostly influenced by its electron, spin, and orbital motions, but its electronic structure is primarily determined by its atoms and their spatial arrangement. In materials science research, crystalline materials are the main constituents of solid materials, with ordered and stable structures, various intrinsic properties, rich physical connotation, clear structure-effect relationship, and ease of control. Modern technology has enabled us to achieve the function-oriented structural design, chemical synthesis, and material preparation to obtain materials and devices with the desired application properties. Therefore, we can build a solid theoretical basis for the function-directed design and preparation of crystalline materials by further deepening and improving our understanding of the scientific nature of crystalline materials as well as the material structure-effect relationships and their implications, which can lead to the discovery and preparation of completely artificially designed revolutionary new materials and promote the rapid development of national economy and science and technology.

After several generations of efforts in crystalline materials research, China has achieved exceptional results in structural design and crystal growth of Nonlinear Optical (NLO) crystalline materials and has been leading the international development efforts in this field since the 1980s. Chinese scientists have invented high-performance nonlinear optical crystalline materials such as Beta-Barium Borate (BBO), Lithium Triborate (LBO), potassium Beryllium Borate Fluoride (KBBF), highly magnesium-doped Lithium Niobate crystals (LN), and L-Arginine Phosphate (LAP), which are famously known as made-in-China crystals with independent intellectual property rights. Chinese scientists also proposed the anionic group theory, dielectric superlattice theory, dual matrix structure model, anionic ligand structure model, and crystal growth defect mechanism, all of which have been added to the international knowledge bank of crystal growth theories and laid a solid foundation for the development of the field. In the meanwhile, Chinese scientists has pioneered the research of dielectric superlattices and paved the way for new optoelectronic functional materials and devices by combining dielectric microstructure and modern crystal growth technology from both theoretical and experimental perspectives, which is an achievement that has gained worldwide attention.

With the increasingly intensified international competition in crystalline materials, especially the increasing relevance of various crystalline materials in vital technologies related to national security, it is imperative to carry out basic research on new functional crystalline materials systematically. However, China generally has few new functional materials with independent intellectual property rights and is conducting too many follow-up studies. The dependence of high-tech industries on foreign technologies is greatly affecting the country's overall competitiveness, particularly in national defense and high technology, which are extremely restricted by Western developed countries. Therefore, China should strengthen the research of new crystalline functional materials to reverse the passive situation in the research of new crystalline materials, develop its independent characteristics in the research of crystalline materials within the shortest time possible, achieve breakthroughs in crystalline materials, innovate a number of new materials with independent intellectual property rights, and reach a new height in nonlinear optical crystal materials in particular.

Materials research not just necessitates a solid understanding of materials science but also relies on various disciplines, including chemistry, physics, and information sciences. Chemists have advantages in the design, controllable preparation, structural control and optimization, and physicochemical characterization of materials, especially in the discovery of new chemical compounds, exploiting the source of new materials research; physicists specialize in the study of new phenomena and properties of materials and their mechanisms, playing an irreplaceable role in the discovery and application of new materials; materials scientists who are the key to the development of new materials from preparation to practical use are responsible for the preparation of materials and focus on optimizing material properties, as well as solve key engineering technologies in the process of materials application; information scientists take the results of new materials research and closely integrate them with major national needs and applications, making the practice of materials more possible. The major research plan hereinafter referred to as the Plan "Structural Design and Controllable Preparation of the Function-Directed Crystalline Materials" thereby aims to develop interdisciplinary cooperation among chemistry, physics, materials, and information to discover new materials, propose new theories, develop new methods for materials design and synthesis, implement applications of materials, and closely integrate the results with national social and economic development, thus shortening the research cycles of materials, meeting major national needs, and accelerating the process of comprehensive strength of China's materials research.

To conduct cutting-edge research on crystalline functional materials, we must, on the one hand, study the molecular and spatial structures of materials thoroughly, and on the other hand, extensively study the macroscopic physicochemical properties of materials and combine quantum chemistry and solid-state energy-band theories to study the electronic structures of materials to seek and determine the structural motifs or other factors that play a dominant role in the functional properties of crystalline materials. We can develop performance-adapted models to predict the physicochemical properties of compounds, validate functionally dominant structural motifs, and further modify and optimize structural motifs for the design and regulation of material properties using theoretical simulations.

The Plan, guided by national demand and the function of materials, focuses on exploring the relationships between structures, compositions, and properties of crystalline materials, and proposes new mechanisms and models for new material exploration; the Plan designs and regulates material structures in response to the demand for material functions, and conducts controlled preparation of a batch of new crystalline materials with specific functions, to establish a new materials research theory, new preparation technology, and new material system for independent innovation, and to open up an important source of knowledge innovation and technological innovation. The discovery of the intrinsic relationships between the optical, electrical, magnetic, and composite properties of crystalline materials as well as their spatial and electronic structures will reveal structural motifs that determine the macroscopic functions of crystalline materials and their integration in space, and provide a theoretical basis for the design and preparation of function-directed crystalline materials. Furthermore, practical applications have put forward many new requirements for photoelectric conversion materials, nonlinear optical crystal materials, laser and fluorescent crystal materials, ferroelectric and microwave dielectric materials, etc.

The overall scientific objectives and breakthroughs of this major research Plan are to discover the intrinsic relationships and laws between the optical, electrical, magnetic, and composite properties of crystalline materials and their spatial and electronic structures, to reveal the functional motifs that determine the macroscopic functions of crystalline materials and their integration in space, and to provide a theoretical basis for the design and preparation of function-directed crystalline materials.

Three key scientific issues were raised during the research period: (i) the identification of key structural motifs for the functions and physical properties of crystalline materials; (ii) the relationships and laws of material functions, physical properties, and their microstructures; (iii) the controllability of the design and preparation of crystalline materials based on functional motifs. The research has been conducted in six areas: the study of functional characteristics and structural primitive relationships of crystalline materials, the synergy and structure-effect laws among functional motifs, the calculation, simulation and functional optimization of crystalline materials, new methods for the design, synthesis and controllable preparation of crystalline materials, new methods for the analysis and characterization of microstructures of crystalline materials, and the functions and applications of crystalline materials. We have made important breakthroughs in the following three aspects: (i) development of a functional-motif theory to guide crystalline materials development with optical, electrical, magnetic, and composite functions, thus creating a new disciplinary growth point at the intersection of physics, chemistry, and materials science; (ii) establishment and development of research methods for controlled crystalline materials synthesis and assembly, detection and characterization of functional motifs, and simulation and prediction of material properties based on the functional-motif theory; (iii) acquisition of a number of crystalline materials that have international influence on and a leading position in related technologies and industries. In particular, we have obtained a series of high-performance material systems in laser crystal materials and nonlinear optical crystal materials, further enhancing the originality of crystalline materials research in China.

This important research initiative has seen a dramatic improvement in the structural design and controllable fabrication of function-directed crystalline materials, starting with functional motifs. We are leading international research efforts in magnetic molecular materials, ferroelectric molecular materials, functional molecular metal-organic framework (MOF) materials, and nonlinear optical crystal materials. We have made great progress in crystalline transparent ceramic laser materials, energy conversion materials, novel Fe-based superconductors, bionic materials, etc. Meanwhile, we have developed new nonlinear optical theories of single-atom polarizable orbitals, and have been the first to discover a number of new deep-ultraviolet nonlinear optical crystal materials. During the Plan, 4,016 research papers were published, including seven in *Science*, three in *Nature*, and 31 in Nature research journals; 536 invention patent applications were submitted, 308 of which were granted, including eight Patent Cooperation Treaty (PCT) patents; 273 guest lectures were given in China and abroad, including 163 international guest lectures; 10 second-class awards of the state Natural Science Award, two second-class awards of the state Technological Invention Award, and one award in Chemistry from The World Academy of Sciences (TWAS) were received.

In terms of talent cultivation, through this Plan, eight project experts or leaders have been elected as academicians of the Chinese Academy of Sciences, 23 have been awarded the National Science Fund for Distinguished Young Scholars by the NSFC, and six have been awarded the Excellent Youth Fund by the NSFC, creating a research team with strong international competitiveness and influence. A new model of research and collaboration on crystalline materials has been established to create an internationally influential research team for interdisciplinary, interpenetrating, and coordinated research in physics, chemistry, and materials science.

We have maintained and developed our strengths in crystalline materials research during this Plan, and proposed strategies and suggestions, which have been adopted by the NSFC and the Ministry of Science and Technology, laying a solid foundation for future development.

Fuzhou, China

Maochun Hong

Contents

Contributors

Kexin Chen National Natural Science Foundation of China, Beijing, China

Rong Chen National Natural Science Foundation of China, Beijing, China

Xianhui Chen University of Science and Technology of China, Hefei, China

Yongjun Chen National Natural Science Foundation of China, Beijing, China

Xuefeng Fu National Natural Science Foundation of China, Beijing, China

Jie He National Natural Science Foundation of China, Beijing, China

Maochun Hong Fujian Institute of Research on the Structure of Matter, Chinese Academy of Sciences, Fuzhou, China

Baosheng Huang National Natural Science Foundation of China, Beijing, China

Yuliang Li Institute of Chemistry, Chinese Academy of Sciences, Beijing, China

Hongcheng Lu National Natural Science Foundation of China, Beijing, China

Jiyang Wang Shandong University, Jinan, China

Qidong Wang National Natural Science Foundation of China, Beijing, China

Yicheng Wu Tianjin University of Technology, Tianjin, China

Jingjun Xu Nankai University, Tianjin, China

Chunhua Yan Peking University, Beijing, China

Junlin Yang National Natural Science Foundation of China, Beijing, China

Shouzhu Zhang National Natural Science Foundation of China, Beijing, China

Guohong Zou National Natural Science Foundation of China, Beijing, China

Project Overview

Maochun Hong, Yicheng Wu, Chunhua Yan, Yuliang Li, Jiyang Wang, Xianhui Chen, Jingjun Xu, Yongjun Chen, Rong Chen, Baosheng Huang, Xuefeng Fu, Junlin Yang, Shouzhu Zhang, Kexin Chen, Qidong Wang, Jie He, Hongcheng Lu, and Guohong Zou

1 Introduction

The major research plan "Structural Design and Controllable Preparation of the Function-Directed Crystalline Materials" (hereinafter referred to as the Plan) was launched by the National Natural Science Foundation of China during the Eleventh Five-Year Plan period (2006–2010) after a thorough demonstration by scientists from various disciplines, including chemistry, physics, and materials. The NSFC's

M. Hong (✉)
Fujian Institute of Research on the Structure of Matter, Chinese Academy of Sciences, Fuzhou, China
e-mail: hmc@fjirsm.ac.cn

Y. Wu
Tianjin University of Technology, Tianjin, China

C. Yan
Peking University, Beijing, China

Y. Li
Institute of Chemistry, Chinese Academy of Sciences, Beijing, China

J. Wang
Shandong University, Jinan, China

X. Chen
University of Science and Technology of China, Hefei, China

J. Xu
Nankai University, Tianjin, China

Y. Chen · R. Chen · B. Huang · X. Fu · J. Yang · S. Zhang · K. Chen · Q. Wang · J. He · H. Lu · G. Zou
National Natural Science Foundation of China, Beijing, China

© Zhejiang University Press 2023

M. Hong (ed.), *Structural Design and Controllable Preparation of the Function-Directed Crystalline Materials*, Reports of China's Basic Research,
https://doi.org/10.1007/978-981-99-3768-4_1

Department of Chemical Sciences, in collaboration with the Department of Engineering and Material Sciences, kicked off the Plan in 2009 after the initial conception and preparation. This Plan has published guidelines and accepted applications six times (2009–2012, 2014, and 2016) and received 658 applications since the first official publication of guidelines and acceptance of applications in January 2009. After the expert peer review and conference review, 158 projects were officially funded (including 124 fostering projects, 29 key projects, 3 integrated projects, and 2 strategic projects) with 188 million RMB in funding. The projects were related to four NSFC scientific departments, i.e., Chemical Sciences, Engineering and Material Sciences, Mathematical and Physical Sciences, and Information Sciences, and mainly belonged to the first two aforementioned departments.

1.1 Overall Scientific Objectives

The Plan was implemented in two main phases. In the first phase, centered around the key issues in the guidelines, the Plan combined specificities and generalizations to solicit projects from the society while focusing on and encouraging free exploration, as well as taking into account the cultivation of some important directions. During the fostering phase, 658 applications were received, 158 of which were granted funding after the expert peer review and conference review. Based on the evaluation of this phase, the Plan was revised and adjusted in accordance with the development trend of the frontier of crystalline materials, with crystalline materials playing the main role.

In the second phase, the major scientific issues were further refined, and based on the research results from the previous phase, integration was underlined, and funding was increased. In 2012, through nearly three years of efforts, this Plan had made promising progress and breakthroughs in all aspects, including molecular-based functional materials, laser crystal materials and nonlinear optical crystal materials, energy conversion materials in terms of their optical functions, magnetic properties, adsorption and catalysis, energy conversion, superconductivity, ferroelectric, and multiferroic properties. To this end, the expert group intends to develop integrated projects in six areas: molecular magnets, molecular-based ferroelectric and multiferroic materials, molecular-based metal–organic framework (MOF) materials, molecular-based optical functional materials, laser crystal materials and nonlinear optical crystal materials, and energy conversion materials. In 2014, the group summarized the previous projects and proposed three integrated directions, i.e., molecular crystalline materials with magnetoelectric functions, molecular ferroelectric materials, and layered superconducting and thermoelectric materials.

Since the official launch of this Plan, it has followed the general principle of "definite objective, stable support, integration and promotion, and leap-forward development" to conduct innovative research at the intersection of chemistry, materials, and physical sciences. To continuously refine significant scientific concerns and aims while promoting interdisciplinary collaboration, the project team combined top-level design with integrated sublimation. In the process of implementation, the academic expert management and project funding management were combined to achieve the maximum result, i.e., the expert team regularly prepared the implementation plan, funding plan, and project guidelines, hosted project reviews and annual communication meetings, reviewed the progress and completion reports, and selected integrated projects; the management team assisted the expert team in strategic planning, organized academic activities and project reviews, and was responsible for the daily management of applications and funded projects. The project team ensured that targeted studies were conducted around major scientific problems and objectives by issuing clear guidelines and combining goal-oriented efforts and free exploration by scientists; ensured the fairness of project reviews and project funding to choose the outstanding applications by tightening the procedures of review by experts online and offline; ensured the quality of project completion by strengthening academic communication and progress follow-ups.

1.2 Overview of Program Implementation

The guidelines of "relying on experts" "scientific management" "encouraging crossover" and "stimulating innovation" were fully reflected in the organization and implementation of this Plan, characterized by application-oriented, in-depth development, and application of basic research. The project implementation used a management structure that combined the fund management system and the expert academic management system. A steering group and a working group were set up for this Plan to establish an orderly, coordinated, interactive, and restrictive working relationship between different responsibilities. The following are specific organizational characteristics.

(i) The academic function of the steering group was given full play, adhering to the guidelines of "expert planning, top-level design, and scientific guidance". The steering group is composed of seven scientists from various fields, and their main responsibilities include conducting strategic research, overall planning, formulating the Plan's implementation plan, formulating annual funding plans and project guidelines, chairing project review meetings, putting forward funding proposals, reviewing progress and final reports, hosting academic seminars and exchanges, conducting field visits to check the project progress, proposing adjustments to the Plan, preparing annual work reports, taking charge of the mid-term self-assessment process, compiling the self-assessment report and phase implementation reports of the Plan, and preparing the summary report and the strategic research report at the end of the program. The Plan's fundamental, forward-looking, and interdisciplinary research characteristics could be well played out under the guidance of experts in the steering group, and the organic combination of major national needs and limited scientific frontier objectives could be achieved.

(ii) The working group's administrative function is defined as "expert support and daily management". The working group is mainly composed of the NSFC's Science Department and Program Bureau staff related to the Plan and the Bureau of International Cooperation. The main responsibilities include assisting the steering group in scientific planning, academic guidance, and strategic research; assisting the steering group in organizing academic activities, such as project inspections, academic exchanges, self-assessments, and summaries; organizing project reviews; undertaking daily management of funded projects; assisting the Program Bureau in organizing mid-term evaluations and acceptance of the Plan; reporting to the committee on phase implementation reports, summary reports and strategic studies of the Plan, etc. The management policies of NSFC's Plan were implemented through the working group, which ensured the smooth workflows of issuing guidelines, application, evaluation, funding, progress and completion checks, assessments, summaries, etc. Meanwhile, the interaction with the steering group was strengthened by providing it with timely information on project application and implementation, assisting it in organizing academic activities, and ensuring that the steering group has first-hand materials for top-level designing, strategic planning, scientific evaluations, integration and improvement.

(iii) A combination of regular and dynamic management was implemented. During the first three years of the program implementation, application guidelines were published annually, including fostering projects and key projects. Applications must conform to the guidelines and reflect the characteristics of the interdisciplinary research and the contributions to core scientific challenges and the program's overall objectives, to ensure compliance with the "definite objective" requirement. A competitive, incentive-based mechanism was introduced during the project approval and guideline issuing processes to greatly fund the projects with the most innovative ideas and scientific value. All projects must undergo a two-tiered review process, namely, peer review and conference review, to ensure the fairness of the project review. Through the dynamic management of the steering group, we could decide to suspend various types of projects with serious problems in the process of implementation, support ongoing projects with good progress, innovation, and breakthrough prospects through integrated research after the mid-term inspection or phase evaluation of this Plan, and strengthen the consolidation of research objectives through the organization of project clusters.

(iv) Interdisciplinary research and subject collaboration on specific topics were promoted and strengthened. In order to strengthen the exchange of academic ideas and information, promote interdisciplinary collaboration and integration, and form research groups, the steering group held information sessions about application guidelines in several cities in China (Hefei, Changchun, Xi'an, Nanjing, Fuzhou, etc.) before the start of the Plan to guide the organization of interdisciplinary research groups. In the course of the project, annual Major Research Plan academic conferences were held (February 28, 2011 to March 2, 2011, Fuzhou; February 20, 2012 to February 22, 2012, Chongqing), making it obligatory for project leaders to participate in the annual academic exchange events organized by the steering group and working group of the Plan where oral presentations and posters were given. The annual academic exchange provided a platform for each theme group to show their research ideas, foundations, conditions, and results, effectively promoting the intersection of chemistry, biology, medicine, bioinformatics, and other disciplines. The annual academic exchange also invited domestic and foreign experts in the field to give academic presentations on special topics, and invited or recommended experts to apply for the next year's major research projects if their research meets the objectives and requirements of this Plan. We believe that the annual academic exchange was successful to promote interdisciplinary research and development.

(v) Initial innovations were scientifically evaluated and encouraged, and interdisciplinary talents were fostered. The steering group evaluated the implementation of the projects in the Plan through the annual academic seminars and exchange conferences on the projects under research. When selecting integrated projects, the project leaders were required to submit five representative papers, evaluate the innovation, scientific value and domestic and foreign impact of the main research, and vote on the projects with better performance to be compiled. The evaluation criteria were strictly based on this Plan's overall objectives and guidelines. Research works that did not meet the objectives would not allow be included in the results statistics. The Plan attached importance to the integration of project support and talent cultivation. During the project, the cultivation of researchers with interdisciplinary backgrounds was one of the KPIs that helped cultivate and retain a group of young and innovative talents with interdisciplinary research ability.

1.3 Interdisciplinary Efforts

This Plan fully reflects the intersection of multiple disciplines. There are interdisciplinary collaborations between chemistry (including organic synthetic chemistry, analytical chemistry, inorganic chemistry, physical chemistry, and structural chemistry) and materials science, efforts between chemistry and physics, and between materials science and physics, thus promoting the cross-disciplinary collaboration among chemistry, materials science, and physics. These disciplines interpenetrate and complement one another in the general direction of new materials research, thereby not only solving important scientific issues in the preparation and characterization of new materials but also promoting related frontier research in various disciplines, meanwhile achieving technological breakthroughs in materials that have been a bottleneck hampering China's economy and national defense security. This is specifically reflected in the following aspects.

(i) Interdisciplinary research of organic synthetic chemistry and MOF materials. On the one hand, the use of organic synthetic chemistry provides specific bridging organic molecular ligands, which greatly facilitates the design and functional application of MOF materials and gives them a variety of excellent and characteristic material functions (such as adsorption, separation, and photoelectric sensing), thus making them even more practical; on the other hand, MOF materials with special structural properties can be easily used to prepare new catalysts with high activity and stability, which can facilitate the recycling of precious metals and reduce the consumption in the catalytic process, thus promoting the application of MOF materials in the field of organic synthesis and achieving a new interdisciplinary field with organic synthetic chemistry.

(ii) Interdisciplinary research of inorganic chemistry and materials science. The advantage of inorganic chemistry in research related to the creation, composition, and structure of new materials, changes between materials, and chemical reaction mechanisms were fully exploited. Interdisciplinary research was conducted to promote the rapid development of new materials to address the challenges of the rapid development of materials science and the new requirements of human beings for high-performance multifunctional materials.

(iii) Interdisciplinary research of chemistry and physics. The collaborative effects of multiple properties in macroscopic materials or systems generated by synthetic chemistry and their connected surfaces, interfacial systems, and carrier generation and transport are key scientific issues in physics and chemistry. As an important objective in the intersection of chemistry and physics, the Plan has well addressed the key scientific issues of how to obtain molecular materials with stable structure and function, how to produce the unique multi-scale effects of molecular materials, how to study the carrier generation and transport characteristics in molecular materials through the intersection of physics and chemistry, how to better understand the charge transfer and energy conversion, and how to establish and develop the synthesis of various bulk phases and surfaces/interfaces.

(iv) Interdisciplinary research of materials science and physics. From the perspective of materials synthesis, we have developed a method to synthesize intercalated iron-based superconductors in liquid ammonia and obtained superconductivity in $A_xFe_2Se_2$ with T_c above 40 K that solid-phase methods cannot obtain. Based on structural design, the synthesis of $Ba_2Ti_2Fe_2As_4O$ superconductor with a complex composite structure has led to a new direction in exploring superconducting materials. In studying thermoelectric materials, we have synthesized several compounds with excellent thermoelectric properties based on modulating thermoelectric properties by structural design of materials. In addition, we have discovered novel structural types that have the potential to become electron crystals and phonon glasses.

(v) Interdisciplinary research of chemistry, materials science, and physics. Chemists have strengths in materials design, controllable preparation, as well as structural control and optimization, and can thus discover new materials; physicists give full play in the study of new properties and phenomena of materials and their mechanisms and thus play an irreplaceable role in discovering new applications for materials; materials scientists focus on optimizing materials properties and solving engineering problems in the process of materials application; the interdisciplinary research of the three disciplines has accelerated the process of meeting the national demand for key high-performance materials in national defense and economic advancement.

2 Research Circumstances

2.1 *Overall Scientific Objectives*

The Plan focuses on discovering the internal relationship between the optical, electrical, magnetic, and composite properties of crystalline materials and their spatial and electronic structures and revealing the functional motifs that determine the macroscopic functions of crystalline materials and their integration in space. It provides a theoretical basis for the design and preparation of function-oriented crystalline materials, and leads the intersection, integration, and development of related disciplines. Important breakthroughs are expected to be made in the following areas.

(i) To develop the functional-motif theory and guide the development of crystalline materials with optical, electrical, magnetic, and other composite functions, thus creating a new disciplinary growth point at the intersection of physics, chemistry, and materials science.
(ii) To establish and develop methods for the controlled synthesis and assembly of crystalline materials, detection and characterization of functional motifs, and simulation and prediction of material properties based on the functional motif theory.
(iii) To obtain several internationally influential crystalline materials that play a leading role in related technologies and industries, especially to obtain one to two high-performance material systems in laser crystals materials and nonlinear optical (NLO) crystal materials, and further enhance the originality of China's crystalline materials research.
(iv) To establish a new research and collaboration pattern for crystalline materials and create an internationally influential research team that is interdisciplinary, interpenetrated and coordinated by related disciplines of physics, chemistry, and materials science.

2.2 *Key Scientific Issues*

The most fundamental issues in the research of function-oriented crystalline materials are discovering solid-state compounds with specific physical properties or functions and investigating their mechanisms of action. This Plan aims to design and synthesize novel solid-state compounds, understand the relationship between their structure and properties/functions, achieve structural regulation of materials, optimize their properties, and improve application performance. The project team will combine forces of scientists from various disciplines, including chemistry, physics, and materials, with the plan to address the following key scientific issues.

(1) Identification of key functional motifs that determine the functional and physical properties of crystalline materials

Although we understand the structure–function relationships, the synthesis and screening of functional materials are often performed randomly. The scientic Plan remain unsolved that is how to design and synthesize new crystalline materials based on the objective laws between structure and function and how to extend and enrich them. For example, high-temperature superconducting materials mostly have a 2D layered structure, such as the CuO_2 structural layer in copper oxides, the graphite structural layer in MgB_2, and the CdI_2 structural layer in Na_xCoO_2. Many solid electrolytes and electrode materials often have a skeletal structure with the disordered distribution of metal-ion carriers. In some solid compounds, it has been found that certain specific structural units may have specific physical properties, i.e., the concept of "functional motifs". For example, thermoelectric materials require both high electrical conductivity and low thermal conductivity, which are physically contradictory. However, if different "functional motifs" are "designed and assembled" in the same crystal, it is possible to achieve both properties in one material. The oxide layer of the CdI_2 configuration in the incommensurate structure compound $[Pb_{0.7}A_{0.4}Sr_{1.9}O_3][CoO_2]_{1.8}$ features high electrical conductivity, while the oxide layer in the NaCl structure features low thermal conductivity. In addition, the incommensurate structure also prevents the effective propagation of thermal vibration waves in crystalline materials; therefore, the reasonable combination of structural units with different properties is an effective way to design and synthesize new crystalline materials with specific functions. To solve this difficult scientific problem, we need to start with a function-oriented structural design focusing on the following two aspects.

(i) Synthesize and study materials and systems with special structures and morphologies, such as cages, spheres, large rings, long chains, micropores, intercalations, lattices, cavities, surface-ordered structures, and the assembly of these functional motifs. The special properties resulting from these structures, such as nonlinear, laser, luminescent, semiconductor, magnetic, composite properties, should also be studied. The structure–property relationships at the atomic and molecular levels should be studied to discover and apply several new functionally sensitive crystalline materials.
(ii) Systematically work on the assembly, modification, and regulation of the photoelectromagnetic properties of crystalline materials and study the synthesis and regulation of their structures and photoelectromagnetic properties; explore the basic scientific problems of long-range electron transport, long-range magnetic ordering, and energy conversion in crystalline materials.

(2) Understanding of the relationships and laws between the functional and physical properties of crystalline materials and their microstructures

The microstructures and symmetry of crystalline materials determine their physical properties. For example, most photoelectric crystals are covalently bonded crystals with an oxide base element and a typical perovskite structure characterized by complex cell structures, narrow energy bands, wide energy gaps, and non-negligible electronic correlation states. These materials have many remarkable physical properties the existing band theory cannot completely explain, such as high-temperature superconductivity, colossal magnetoresistance, anomalous piezoelectric and ferroelectric behavior, and impurity sensitivity of phase transitions. On the one hand, this heralds a breakthrough in band theory, which is the mainstream theory of contemporary materials science. On the other hand, it opens up opportunities to study functional crystals and their microstructures. To explore this scientific challenge, we must focus on the following three aspects.

(i) Establish and develop new theoretical methods to calculate, simulate, and predict the structures and properties (e.g., magnetic, electrical, and optical) of materials at multiple levels and scales and to design crystalline materials with specific functions and significant practical contexts. Collaborate closely with other related Plan to provide theoretical support for structural and performance studies of new materials.
(ii) Reveal and understand the relationship between the structures (e.g., atoms, ions, molecules, groups, etc.) of functional motifs, their bonds (e.g., covalent, ionic, ligand and hydrogen bonds, π-π stacking, etc.) and their properties (e.g., optical, electrical, magnetic and complex). Reveal the relationship between the macroscopic symmetry of crystalline functional materials and their functional properties.
(iii) Investigate the physical response and regulation of the relevant systems under external perturbations (e.g., magnetic field, electric field, optical field, temperature field, pressure, etc.) and find control techniques of practical value. Explore multiferroic crystalline materials with strong magnetic and ferroelectric properties and find new preparation methods.

The gradual resolution of these problems will deepen our understanding of the fundamental processes of light, electricity, and magnetism and the interrelationships between the atomic and molecular (group) structures, symmetry, and energy-band structures of crystals, thus providing a solider theoretical basis for crystal design.

(3) Research on design principles and controllable preparation of crystalline materials based on functional motifs

Through the structure-effect relationships and laws of crystalline materials discovered earlier, we can design structures based on functional requirements and design synthesis reactions based on structures, which achieves controlled synthesis and assembly of crystalline materials, thus exploring new physical effects and discovering new materials. In addition, the discovery of new crystalline materials often requires new synthesis methods and techniques, but turning them into functional materials with practical applications requires in-depth testing, analysis, and evaluation of their physical properties and microscopic mechanisms, which further requires the development of new methods for material characterization. Therefore, according to the development trend of crystalline functional materials, this Plan will foster close collaboration between chemists, physicists, and materials scientists to develop and refine new methods for the synthesis, preparation, and characterization of crystalline functional materials. The core scientific problems are reflected in the following five areas.

(i) Modify or dope of selected crystalline functional material systems to achieve controlled growth of crystal structures and morphologies, development of functional-motif assembly methods, and functional enhancement and compounding of crystalline materials through structural regulation.
(ii) Establish new methods for detecting and characterizing functional motifs and materials, focusing on developing in-situ, time-discretized, and micro-area structure characterization techniques to comprehensively measure the relevant properties of synthetic materials.
(iii) Conduct systematic research on the assembly technology of crystalline materials to prepare new functional crystalline materials through structural design and tailoring of functional motifs. We have investigated the assembly mechanism, the driving force of the assembly process, the controlling factors, and the interrelationship between functional motifs and macroscopic functions.
(iv) Develop new methods for the synthesis and characterization of unconventional crystalline materials with a focus on new techniques for synthesizing sub-stable crystalline materials under extreme conditions, such as new preparation methods of thin-film materials, high-temperature and high-pressure synthesis, and soft chemical preparation methods.
(v) Develop new theories, methods, and techniques with high spatial, energy, and time resolutions based on China's large-scale scientific facilities to provide microstructural information to study the physical properties and mechanisms of crystalline materials.

3 Significant Progress

Taking full advantage of interdisciplinary cooperation among chemistry, materials science, and physics, this Plan is oriented toward the optical, electrical, magnetic, and composite functions of crystalline materials, and conducts in-depth research on the most fundamental "compositional and structural basis" of crystalline materials, i.e., functional motifs, reveal those motifs that determine the macroscopic functions of crystalline materials and their spatial integration approaches, and provide a theoretical basis for the structural designing and controllable preparation of function-oriented crystalline materials. We have made significant progress in the following directions and achieved leapfrog development.

(1) Taking the lead in molecular-based materials research worldwide

   (i) Magnetic molecular materials. We were the first in the world to propose a symmetrical strategy for the construction of rare-earth single-molecule magnets (SMMs) and report the first case of organometallic single-ion magnets. A new record for blocking temperature of SMMs was created and maintained. The research work we have done in this area has taken the lead across the world.
   (ii) Ferroelectric molecular materials. We were the first to observe ferroelectric domains and discover the orderly coexistence of ferromagnetism and ferroelectricity. We were also the first to propose a new theory of symmetry breaking to produce ferroelectric magnetic-electric coupling materials.
   (iii) Functional molecular MOF materials. Based on molecular design and crystal engineering principles, we have discovered novel porous structures and dynamically changing behaviors and achieved excellent adsorption, separation, sensing, catalysis, and electrical conductivity functions.

(2) Holding the leading position in NLO crystal materials research worldwide

   (i) We proposed a new structural design for complex asymmetric structure functional groups to produce the NLO effect, developed the NLO crystal theory, and thus were the first to discover several new functional crystalline materials globally. The research on NLO crystals plays a leading role in this field internationally.
   (ii) We proposed the theory of "synergy of two asymmetric structural motifs" and experimentally confirmed the enhanced synergy of the dipoles of asymmetric monomers, which led to the discovery of a series of novel antimony-sulfur infrared NLO crystals.
   (iii) We proposed the new theory of grain boundary engineering and polycrystalline microstructure, which has brought a breakthrough in crystalline transparent ceramic laser materials, making China the second country after the United States to achieve multi-sheet stacked 10,000-W laser output.

(3) Achieving significant breakthroughs in energy conversion materials research

   (i) Through the synergy and controlled synthesis of multifunctional motifs, we made a breakthrough in the research of novel thermoelectric materials. We discovered a series of new antimony-based thermoelectric materials with excellent properties, thus taking the lead in this field.
   (ii) The discovery of a series of novel Fe-based superconductors was reported by *Materials Research Society* and *American Physical Society*, which has led more than 350 research groups in 41 countries to carry out follow-up research, making a significant impact internationally.

(4) Promoting the development of chemistry, materials science, and physics in China

We have provided a solid and effective platform for cooperation and exchange among the disciplines of chemistry, materials science, and physics in China and achieved significant research results with interdisciplinary significance.

(i) We were the first to synthesize chalcogenide oxide (ABO_3) with triple-valence Mn, which had a single crystal with a p–n junction at the atomic or molecular scale, opening up a new direction in the regulation of the properties of crystalline materials.
(ii) The hybridization of long-range ordered nano semiconductors with a high surface area prepared by electrochemical methods with organic light-absorbing materials effectively improved the interfacial contact between electron-transporting and hole-transporting materials and also increased the light-absorbing efficiency and carrier mobility, leading to the development of a series of novel structures for fiber cells.
(iii) We developed the design and chemical control of inorganic phase change systems with one-dimensional atomic chain structures, revealed the structural evolution of these systems, and experimentally verified the prediction of the famous solid-state chemist Goodenough in the 1970s that the atomic spacing in the atomic chain would affect the electrical properties, and proposed a representative method to obtain the hydrogenated VO_2 structure.
(iv) We were the first to report the multiple charge-ordered states and nano-phase separation in $LuFe_2O_4$ and to study the strong nonlinear electrical transport properties in $LuFe_2O_4$ systematically; we were also the first to report the nano-polarization domains and strong magnetoelectric coupling effects in the Fe_2OBO_3 charge-ordered system and to provide a correlation model between them.
(v) We combined first-principle calculations with synchrotron radiation and neutron diffraction. We made great research progress in characterizing and understanding new functional materials' magnetic and electrical properties and coupling effects.

The implementation of this Plan has greatly enhanced the innovative capabilities of our multidisciplinary research teams in chemistry, materials science, and physics. Scientists have taken the initiative to cultivate their sense of innovation when conducting research and closely focused on studying the functions of crystalline materials and their relationship with their structures, exploring the synergy and structure-effect laws between functional units, and enabling the structural designing and controllable preparation of crystalline materials. Through the integration of multiple disciplines and several important original innovations, world-leading innovative research results have been achieved in the following important research areas.

(i) We have designed and synthesized a series of SMMs from functional motifs and opened up a new field of single-ion magnets with a record-breaking energy barrier (>2,200 K), and their magnetic blocking temperatures above the liquid nitrogen temperature range.

(ii) We proposed to use space-confined supramolecular interaction to enhance molecular recognition and challenged the limitation of traditional porous materials to adsorb olefins preferentially; we synthesized microporous MAF (mobile amorphous fraction) crystalline materials with reverse adsorption selectivity for the purification of butadiene.

(iii) We were the first to discover the effect of hydroxyl radicals on molecular sieve crystallization. We proposed a new mechanism for generating molecular sieve crystalline materials, which was an important breakthrough in the mechanism study.

(iv) We designed and synthesized an ultrathin 2D crystalline material whose unique electronic structure can greatly reduce the activation energy barrier of CO_2, thus achieving efficient electro-reduction of CO_2 to liquid fuel.

(v) We developed a new method of "phase interface strain" to regulate the lattice strain by two-phase eutectic to produce giant polarized crystalline ferroelectric thin-film materials, which increased the polarization of classical $PbTiO_3$ ferroelectrics by a factor of 3.8 and increased the highest value of ferroelectrics previously reported internationally by 80%.

(vi) We were the first to prepare an artificial nacre crystalline material similar to natural pearl layers through biomimetic mineralization preparation. The crystalline material produced by this method is strong and ductile and can be used to prepare bone grafts with nacre-like structures, as well as a variety of new biomimetic engineering materials. The method was described as "a breakthrough" by Perspectives in *Science*.

(vii) We were the first to observe ferroelectric domains and discover the orderly coexistence of ferromagnetism and ferroelectricity. We were the first to propose a new symmetry-breaking theory to produce ferroelectric magnetoelectric coupling materials. As the pioneer in metal-free ABX_3-type chalcogenide ferroelectric molecules for practical applications of molecular ferroelectrics, we designed and synthesized the world's first molecular ferroelectric with enantiomers.

(viii) We developed a new NLO theory of single-atom polarizable orbitals and were the first to discover a new class of deep-ultraviolet (deep-UV) NLO crystal materials. We proposed the theory of "synergy of two asymmetric structural motifs" which led to the discovery of a series of mid and far-infrared NLO crystal materials. A new theory of grain boundary microstructure was proposed, which made a breakthrough in crystalline ceramic laser materials, making China the second country after the United States to achieve a 10,000-W laser output. We have once again led the research in NLO crystals in the international arena and have made milestone progress in studying NLO theory.

(ix) We discovered the new KFe_2Se_2 series of iron-selenium-based superconductor, the first internationally recognized new superconductor discovered by

Chinese scientists, thereby opening up a new field of international superconductivity research and keeping China's iron-based superconductivity research in a leading position in the international arena.

During the project implementation period, 4,016 research papers were published, including 7 in *Science*, 3 in *Nature*, and 31 in *Nature* series; 536 invention patent applications were submitted, 308 of which were granted, including 8 PCT patents; 273 guest lectures were given at home and abroad, including 163 international guest lectures and 110 domestic guest lectures; 10 s-class awards of the National Natural Science Award, 2 s-class awards of the National Technical Invention Award, 1 award in Chemistry from The World Academy of Sciences (TWAS), 18 first-class awards of the Provincial and Ministerial Natural Science Award, 9 s-class awards of the Provincial and Ministerial Natural Science Award, 5 first-class awards of the Provincial and Ministerial Technical Progress Award, and 8 s-class awards of the Provincial and Ministerial Technical Progress Award were received.

With regard to talent cultivation, through the implementation of this Plan, 8 project experts or leaders have been elected as academicians of the Chinese Academy of Sciences, and 23 youths have been awarded by the National Science Foundation for Distinguished Young Scholars, which works together as a research team with strong international competitiveness and influence.

Table 1 compares the development trends in various fields before and after the Plan was completed.

Table 1 Comparison of development trends in various fields before and after the completion of the major research plan "Structural Design and Controllable Preparation of the Function-Directed Crystalline Materials"

Key scientific issues	Research status in China at the start of the plan	Research status in China at the end of the plan	International research status at the end of the plan	Strengths and gaps compared with international research status
How to design and regulate the structures of magnetic functional molecular crystalline materials	Domestic research on magnetic molecular materials was still in its infancy, and there was no clear theory or strategy to enable the performance regulation of single-molecule magnets	Established an innovative idea of symmetry regulation and obtained the first metal–organic molecular magnet with the highest blocking temperature and effective energy barrier in the world	The current research focuses on the synthesis of high critical temperature molecular magnets, high blocking temperature single-molecule magnets and 4d/5d heavy transition system single-chain magnets, the exploration of molecular magnetic materials for spin manipulation, information storage and other applications, the establishment of corresponding characterization and evaluation tools, and the construction of molecular spintronic devices with practical applications	The research is among the leading ones in the world

(continued)

Table 1 (continued)

Key scientific issues	Research status in China at the start of the plan	Research status in China at the end of the plan	International research status at the end of the plan	Strengths and gaps compared with international research status
How to design and synthesize high performance molecular ferroelectrics and regulate their properties	The research on molecular ferroelectrics was at the starting stage with a weak research base	Discovered the first metal-free chalcogenide, the highest polarization strength molecular ferroelectric, and the highest piezoelectricity of molecular materials	The current research focuses on developing universal strategies for building multipolar axial molecular ferroelectrics, exploring the modulation of ferroelectric properties and other functional optimizations, and preparing molecular ferroelectric films and realize their device applications	Leading the research internationally, developing with multiple focuses and entering the era of chemical design
How to develop new laser crystals and NLO crystals and explore the relationships between their structures and functions	Leading the basic research on laser crystals and NLO crystals in the world, but with unbalanced development of practical applications	Developed single-atom response methods and explored new systems for large-diameter titanium gemstones, laser self-multiplying crystals, and deep-UV NLO crystals	The current research focuses on establishing the energy-band structure model of phonon-photon coupling in mid- and far-infrared nonlinear crystals to obtain new NLO crystals with high damage and high efficiency in the mid- and far-infrared; laid the material foundation for the integration of active optical quantum chip devices through precise regulation of optical superlattice microstructures in the communication band; developed new methods for microscopic regulation of structural primitives to achieve refractive index regulation in the solar-blind band and efficient laser performance in this band performance output	Maintained our leadership in basic research and continued to make breakthroughs in practical applications

Research in China and Abroad

Maochun Hong, Yicheng Wu, Chunhua Yan, Yuliang Li, Jiyang Wang, Xianhui Chen, Jingjun Xu, Yongjun Chen, Rong Chen, Baosheng Huang, Xuefeng Fu, Junlin Yang, Shouzhu Zhang, Kexin Chen, Qidong Wang, Jie He, Hongcheng Lu, and Guohong Zou

Materials form the basis of human material civilization and serve as a forerunner to the national economy, social growth, and national security. Materials science's intersection and integration with information science, environmental and energy science, and life science have become an important trend and driving force for materials development. The materials engineering and the materials genome initiative development has resulted in revolutionary changes in materials science and technology, as well as the emergence of many advanced materials with unique properties. The research and development of advanced materials for information, energy, and life science

M. Hong (✉)
Fujian Institute of Research On the Structure of Matter, Chinese Academy of Sciences, Fuzhou, China
e-mail: hmc@fjirsm.ac.cn

Y. Wu
Tianjin University of Technology, Tianjin, China

C. Yan
Peking University, Beijing, China

Y. Li
Institute of Chemistry, Chinese Academy of Sciences, Beijing, China

J. Wang
Shandong University, Jinan, China

X. Chen
University of Science and Technology of China, Hefei, China

J. Xu
Nankai University, Tianjin, China

Y. Chen · R. Chen · B. Huang · X. Fu · J. Yang · S. Zhang · K. Chen · Q. Wang · J. He · H. Lu · G. Zou
National Natural Science Foundation of China, Beijing, China

© Zhejiang University Press 2023

M. Hong (ed.), *Structural Design and Controllable Preparation of the Function-Directed Crystalline Materials*, Reports of China's Basic Research,
https://doi.org/10.1007/978-981-99-3768-4_2

applications is the hot topic of international competition, such as advanced opto-electronic materials for the information industry, energy conversion and new energy storage materials for the energy industry, new biological and medical materials for the life science. The development of cutting-edge materials should gather resources to focus on top-level designs and systematic research to meet major national needs and seize the commanding heights of competition in the materials industry while paying more attention to interdisciplinary and disruptive technological innovation. In materials science research, crystalline materials, which are characterized by their orderly and stable structures, various intrinsic properties, rich physical connotations, clear structure-effect relationships, and ease of compound control, are the main bodies of solid materials and will enable function-oriented structural designs, chemical synthesis, and material preparation to obtain materials and devices with desired application properties.

1 Research Status and Development Trend

This Plan started in 2009. It has been widely recognized that the source of materials is chemical compounds. The synthesis of chemical compounds is the fundamental task of chemistry, therefore it is necessary to start with chemistry and the design and synthesizing of new compounds to reinforce the originality of materials research. The development of modern science and technology has made it possible to work on chemical synthesis from an inverse process, i.e., to design compounds with specific structures according to specific property requirements using molecular engineering methods, thereby obtaining compounds with the desired properties.

The interdisciplinary research of materials science and physics and the use of advanced physics theory to guide the design, preparation, and application of new concepts of advanced functional materials will generate a series of new materials and devices that can meet the demand for high-performance functional materials in China's future emerging industries, such as 5G/6G, quantum communication and quantum computing, the Internet of Things, human–computer interaction, virtual reality, and autonomous driving. Devices with a high response time, low latency, low power consumption, and high fidelity are required for new applications. The functional materials also develop toward high performance, low cost, integration, compositing, intelligence, and digitalization. In the twenty-first century, due to the development of quantum physics theory and the advancement of precise experimental preparation technology, our understanding of materials has advanced to a microscopic level, enabling us to observe and understand the correlations between materials of various scales and dimensions and the resulting rich derivative and cooperative phenomena. New research fields such as quantum information, quantum computing, artificial band-gap materials, topological physics, condensed topological phases, and topological phase transitions have been developed, which can lead to the preparation of materials with unprecedented special properties, including completely artificial materials. Materials science and physics interdisciplinary research will produce new disruptive materials and devices, enabling China to become one of the international leaders in this field of research, master the key "stranglehold" of future-proof technologies, and develop its original innovation, core competition, and strategic deterrence capabilities to lead the development of science and technology in this field.

Data and artificial intelligence are the key forces behind modern technology and the industrial revolution and are at the cornerstone of research paradigm changes in various fields, such as materials science. Accelerating the development of next-generation big data and artificial intelligence technologies is a strategic issue related to the new technological and industrial revolution opportunities. Artificial intelligence technologies can be used to empower basic research in materials science, break through bottlenecks in materials, mathematics, physics, and chemistry, improve our ability to solve major materials science problems, accelerate research paradigm changes in materials science, and even completely relovotionize materials science research.

1.1 Research Status and Development Trend of Optoelectronic Functional Crystals

Optoelectronic functional crystals exhibit optoelectronic properties and constitute an indispensable material basis for optoelectronic technology. They are classified as laser crystals, NLO crystals, piezoelectric crystals, and scintillation crystals based on their functional properties. China has developed internationally influential theoretical models and material bodies such as the "anion group theory" and created an ideal foundation and conditions for accelerating the development of optoelectronic functional crystal theories and the exploration and industrialization of new crystals, making her an international leader in the exploration, growth, and application of new functional crystals represented by inorganic NLO crystals. Currently, optoelectronic functional crystals are evolving toward highquality, largesize, low dimensionality, compositing, functional integration, and miniaturization to meet the requirements of optoelectronic devices represented by all-solid-state lasers for use in extended wavelengths, high frequencies, short pulses, and complex extreme conditions. For example, the need for materials can withstand harsh and complex environments for a long time, the need to obtain some crystalline materials with special functional properties in extended (new) wavelengths such as mid-and far-infrared and sensitive wavelengths, and to focus on the application of the functional crystals in high power and complex conditions, all of which demand higher requirements for optoelectronic functional crystals. New optoelectronic functional crystals are being developed to meet the needs of national economic and social development as well as national defense and security.

1.2 Research Status and Development Trend of Molecular Ferroelectrics

In recent years, researchers in China have made several advancements in molecular ferroelectrics. In 2013, the research group of Rengen Xiong discovered that the Curie temperature (T_c) of diisopropylamium bromide (DIPAB) was 426 K. The saturation polarization (P_s) was 23 $\mu C/cm^2$, at a point the T_c and P_s of this molecular ferroelectric approached the level of $BaTiO_3$ (Barium titanate) for the first time [1]. Subsequently, in 2017, the same group designed and synthesized molecular ferroelectric materials TMCM-$MnCl_3$ [trimethyl chloromethyl ammonium-manganese (3^+) chloride] and TMCM-$CdCl_3$ (trimethyl chloromethyl ammonium-trichlorocadmate) with a piezoelectric coefficient (d_{33}) of 185 pC/N and 220 pC/N, respectively [2]. These values outperform all previous molecular materials while approaching BaTiO3 levels. Then, in 2018, they fabricated the first metal-free organic perovskite ferroelectric [3], introducing a new member to the important perovskite materials family. Recently, the group discovered the first molecular ferroelectric solid solution with a morphotropic phase boundary (MPB) [4] and showed soft molecular ferroelectrics with piezoelectric properties comparable to those of hard commercial inorganic ferroelectric ceramics. This group was also the first to discover the organic enantiomeric ferroelectrics [5]. Furthermore, Rengen Xiong's group has developed new approaches such as the "Quasi-Spherical Theory" "introducing homochirality" and "hydrogen/fluorine substitution" to design molecular ferroelectrics [6]. These methods enable the chemical design of molecular ferroelectrics without relying on crystal databases or blind searching. After years of hard work, Chinese scholars have progressed from followers to neck-to-neck runners and eventually to leaders in the international molecular ferroelectrics industry.

1.3 Research Status and Development Trend of Molecular Magnetics

Single-molecule magnets (SMMs), unlike macroscopic magnets, are single-domain nano-quantum magnets with homogeneous size and composition. The direction of the magnetic moment can be maintained for a long time below the magnetic blocking temperature without flipping. In addition, SMMs also show other unique and interesting quantum behavior. Therefore, they have great potential in applications for high-density information storage, quantum computing, and spintronics devices. Since their discovery, increasing the effective potential energy barrier (U_{eff}) and magnetic blocking temperature (T_b) has been a core goal in designing and synthesizing high-performance SMMs. Numerous advances in unimolecular magnets have been made since 1993 when the Gatteschi group from Italy first discovered $\{Mn_{12}\}$ showing unimolecular slow magnetic relaxation [7].

In transition-metal SMMs, researchers have discovered the "star" SMMs of transition-metal clusters such as Mn_3 and Fe_8. Detailed studies of the formation mechanisms of SMMs showed that large ground-state spin values and uniaxial magnetic anisotropy are critical to forming single-molecule magnets. Subsequently, they discovered that even the synthesis of Mn_{19} with four times the ground-state spin value of Mn_{12} (ground-state spin value $S_T = 83/2$) [8] was not effective in increasing U_{eff}.

Regarding rare-earth SMMs, in 2003, the Ishikawa group from Japan, discovered that double phthalocyanine-sandwiched rare-earth ionic compounds have sluggish relaxation behavior even under zero DC field [9], which was attributed to single-ion behavior by magnetic dilution. This is the first and most studied single-molecule magnet with a single rare-earth ion. In 2008, the Coronado group from Spain discovered that a single Er^{3+} ion exhibited similar slow magnetic relaxation characteristics in a tungsten polyacid (WA) system but without hysteresis behavior [10]. In 2011, the first rare-earth organometallic monotonic magnet COTLnCp* was fabricated by the Song Gao research group from China using cyclooctatetraene anions (COT^{2-}) and metallocene derivatives (1, 2, 3, 4, 5-pentamethylcyclopentadienyl, Cp^{*-}) with heavy rare-earth ions Dy^{3+}, Ho^{3+}, and Er^{3+} [11], thus paving the way for rare-earth organometallic monotonic magnets. In 2013, Mingliang Tong's group proposed a symmetry strategy and discovered rare-earth SMMs with their T_b up to 20 K and U_{eff} exceeding 1000 K [12–14]. Later on, the groups of Layfield and Mills [15, 16] in the UK discovered rare-earth SMMs with T_b up to 60 K. Recently, the group of Mingliang Tong, in collaboration with Layfield's group, first obtained a T_b surpassing the liquid nitrogen temperature range [17]. In the next decade, the study of magnetic quantum materials and devices based on molecular systems will be a new direction of quantum-information and quantum-computing research. It will become a popular research field worldwide [18, 19].

1.4 Research Status and Development Trend of Superconducting Materials

As a new quantum material type with zero resistance and complete diamagnetism, superconducting materials have profound physical implications and wide application prospects. Conventional superconductors are currently used in many applications, including energy, medical, sensing, communication, quantum computing, and particle acceleration. Such materials' transition temperature, critical current density, and critical magnetic field, however, are still relatively low. For superconducting technology to be commercialized on a large scale, research toward high-performance, high-temperature superconducting materials is required. Furthermore, studies of the mechanisms behind these materials' unconventional high-temperature superconducting behavior are of great significance for understanding complex quantum phenomena and developing cutting-edge technologies and theories. Recently, after the discovery of copper-based superconductors, the successive discovery of iron-arsenic-based and iron-selenium-based high-temperature superconducting materials has provided a brand-new opportunity for both fundamental and applied research on unconventional high-temperature superconductors.

Recently, several advances have been made in high-temperature superconducting materials. In the field of unconventional superconductors, Hosono's group from Japan discovered iron-arsenic-based high-temperature superconductors in 2008, after the discovery of copper-oxide high-temperature superconductors [20]. Subsequently, the T_c for iron-arsenic-based superconductors manufactured by Zhongxian Zhao's group in China reached a record high of 56 K [21]. In 2010, Xiaolong Chen et al. from China developed a new iron-selenium-based high-temperature superconductor, KFe_2Se_2 (potassium-doped iron selenide), which attracted the attention of many scientists [22]. Thereafter, Xianhui Chen et al. discovered LiOHFeSe (lithium iron-selenide hydroxide) with a T_c of up to 40 K [23], and Qikun Xue's group identified the interface-enhanced high-temperature superconductivity of monolayer FeSe (iron selenide) (with a T_c up to 65–100 K) [24]. For unconventional superconductors, more extensive and comprehensive measurements on CuO (copper oxide) and Fe-based materials are expected in the future. Experimentally, disruptive results may be achieved through more direct measurements, while theoretical studies must be further developed to explain key experimental phenomena completely.

As for practicality, Yanwei Ma's group from China have developed iron-based high-temperature superconducting wires with their high field critical currents reaching practical levels. The technology breakthrough indicates that these materials have broad application prospects in high critical field currents, ease of iron-based superconductors processing, and practical wires.

In the field of conventional superconductors, significant advances have been made in the study of BCS (Bardeen-Cooper-Schrieffer) superconductors at high pressures. Inspired by the theoretical calculations of Yanming Ma's group, the research group of Eremets from Germany was the first to apply a high pressure of 150 GPa to H_2S in 2015 and obtained superconductivity with a T_c of 203 K [25]. Recently, Eremets' group. further observed that LaH_{10} has superconducting behavior up to 250 K at a high pressure of 170 GPa [26]. The superconductivity of such hydrogen-containing compounds is close to room temperature (−20 °C) but still requires extreme high-pressure conditions. Therefore it is not practical for many applications. In the study of high-pressure hydrides, Chinese scientists have predicted numerous high-pressure superconducting phases containing hydrogen, and more systems are expected to be discovered. It shows greatly challenging to study these systems regarding microscopic mechanisms and physical properties. Theoretically, these materials may be BCS superconductors at high phonon frequencies, which needs further experimental data to confirm. More in-depth studies of such systems will be performed in the future.

1.5 Research Status and Development Trend of Thermoelectric Materials

As a new type of clean energy materials, thermoelectric materials can directly achieve interconversion between thermal and electrical energy, potentially providing a feasible option for improving energy efficiency and alleviating environmental pollution [27]. However, the inverse coupling relationship between electrical and thermal transport parameters determined by the intrinsic electronic structure of the material limits the figure-of-merit (ZT value) of thermoelectric materials and the energy conversion efficiency of thermoelectric devices, which has prevented them from being widely applied. Recently, great progress has been made internationally in thermoelectric materials in moderate and high-temperature zones (600–900 K). Using multi-scale defect modulation and energy-band modulation, research groups at Northwestern and other US universities have significantly improved the thermoelectric performance of p-type thermoelectric materials, such as PbTe (lead telluride) with a ZT value of 2.3–2.5 and SnSe (tin selenide) with a ZT value exceeding 2.5 [28]. However, the temperature of most industrial and domestic waste heat is considered as low grade (below 300 °C), and the efficiency of thermoelectric materials for power generation at such temperatures is under 4% due to slow advancement in this area. Therefore, there is an urgent need to develop high-efficiency thermoelectric materials and devices for low-temperature operations. Scientists in the United States and China have developed $Bi_{2-x}Sb_xTe$-based bulk and thin-film thermoelectric materials with a ZT value of 1.4–1.5 at low temperatures using nanostructuring and energy-band modulation [29]. In addition, new low-cost MgAgSb alloy thermoelectric materials have been developed with a ZT value of 1.2–1.4 that is still too low to be practical.

2 Development Trends of Functional Crystalline Materials

The functions of materials are derived from their optical, electrical, magnetic, pressure, and mechanical properties or combinations thereof (e.g., photovoltaic, electro-optical, acoustic-optical, magnetoelectric, and thermoelectric effects) and are evolving from single-functional materials to multifunctional composites. The photoelectromagnetic properties of a material depend mainly on its electron spin and orbital motion, while its electronic structure depends mainly on the constituting atoms and their spatial arrangement. This Plan aims to uncover the internal relationships between the optical, electrical, magnetic, and composite properties of crystalline materials and their spatial and electronic structures, to reveal the structural elements that determine the crystalline materials' macroscopic functions and their integration in space, to provide a theoretical basis for the design and preparation of function-oriented crystalline materials, and to establish a new theory of materials research, new preparation technology, and a new material system with Chinese characteristics.

In this Plan, the project team performed in-depth research on the molecular and spatial structures and macroscopic properties of materials and studied the electronic energy band and magnetic structures of materials in conjunction with related advanced theories. The team also investigated and identified the structural motifs that play a dominant role in the functional properties of crystalline materials, explore their conformational relationships, established property-adapted models, designed and predicted the properties of compounds, verified the role of functional motifs, and achieved the design and control of the material properties with further modification and optimization. Finally, it designed and fabricated advanced laser crystals and NLO crystals, high-performance molecular ferroelectrics, and molecular magnets layered superconducting, thermoelectric, and MOF materials. Although the materials were at different stages of development when this Plan was initiated, through nearly a decade, all research on the aforementioned functional materials have taken the lead and are of significant impact on attracting international attention (Table 2).

2.1 Developments in Laser Crystals and Nonlinear Optical Crystals

In the field of laser crystals and NLO crystals, researchers have developed a new NLO model with single-atom polarizable orbitals, which takes important progress since we proposed the far-reaching "anion group theory" in the international arena. In addition, we were the first to discover new deep-UV NLO crystal materials and keep leading the basic research in this field. From 2008 to 2019, the number of papers published in the field of NLO crystal materials in China accounted for an increasing proportion of the global number of papers (45% in 2019), while the annual number of papers published in the US has been hovering around 13%; the number of ESI highly cited papers in China was 41, ranking first in the world (Fig. 1).

In addition to the leading achievements in applications of deep-UV NLO crystals applications and the large-size titanium-doped sapphire with the highest 10 PW in the world, China has also developed a new 1.55 mm eye-safe Er:Yb:YAB crystal in this field, which can be used for automatic piloting and carrier landing guidance lidar and optical communication amplifiers, thus continuing to solve the stranglehold issue and enhance China's leading position in this field. Meanwhile, we also proposed the "synergy of two asymmetric structural motifs" which led to the discovery of a series of mid-and far-infrared NLO crystal materials urgently needed for major national applications. For practical applications, our leading laser self-multiplying crystals have been applied to green light modules in the fields of distance measurement and anti-terrorism, and we have proposed a new mechanism of phonon modulation energy level to extend the laser wavelength to cyan and yellow wavelengths to meet the major needs of medicine and navigation in fog. Guided by the new theory of grain boundary microstructure, a breakthrough has been made in crystalline ceramic

Table 2 Research content, project progress, and highlights of this plan

Material type	Status at the start of the Plan	Status at the end of the Plan	Highlights
Laser crystals and NLO crystals	Leading the basic research worldwide with unbalalnced application development	Leading in basic research with multiple breakthroughs in practical applications	Large-diameter titanium gemstones, human-eye safety, new systems of laser self-multiplying crystals and deep-UV crystals, single-atom response methods
MOF materials	Hot topic, follower	Received international attention with outstanding highlights	Efficient separation of low-carbon olefins, and high CO_2 adsorption materials that are among the best worldwide
Molecular ferroelectrics	Started initial research, and followed up	Leading the research worldwide with multi-focus development, and entering the era of chemical design	The first metal-free chalcogenide, the highest polarization strength molecular ferroelectric, and the highest piezoelectricity of molecular materials
Magnetic functional molecular materials	Started initial research with confusion	Innovative idea of symmetry regulation, with a leading material system worldwide	World's first metal–organic single-ion magnet worldwide with the highest blocking temperature and effective energy barrier
Layered superconductors	Hot topic, co-leader	Outstanding highlights, with world-class innovations	Discovery of layered iron-selenium-based superconducting materials, leading international superconductivity research
Thermoelectric materials	Hot topic, co-leader	Received international attention with outstanding highlights	Defects and dimensional regulation, a new way to improve thermoelectric properties

laser materials, making China the second country after the United States to achieve multi-sheet stacked 10,000 W laser output.

2.2 *Development Trend of Molecular Ferroelectrics*

In the past decade, molecular ferroelectrics have become the most popular materials due to their chiral introduction, ease of shearing, good mechanical flexibility,

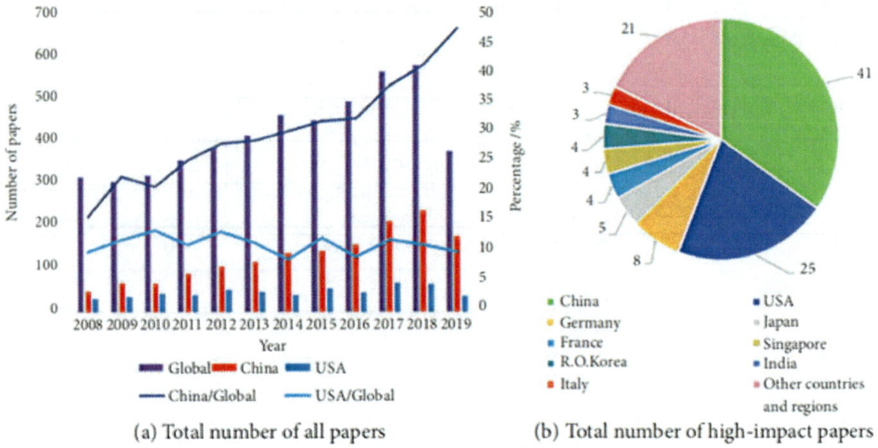

Fig. 1 Global papers on NLO crystal materials from 2008 to 2019

good biocompatibility, acoustic impedance matching with the human body, environmental friendliness, low cost, and lightweight. They are expected to be a useful complement to inorganic ferroelectrics because of their aforementioned advantages. Through chemical design and controlled synthesis, molecular ferroelectric devices can be developed for various foldable, biocompatible, or wearable applications. More importantly, most of the film production processes using molecular ferroelectrics are simple with mild conditions, which can greatly reduce the energy consumption and cost of the corresponding devices. This facilitates the large-scale application and promotion of ferroelectric random-access memory, flexible piezoelectric devices, wearable sensors, etc. The research efforts of China in the field of molecular ferroelectrics can be seen in Fig. 2.

2.3 Development Trend of Molecular Magnets

Since the implementation of this Plan in 2008, the research on magnetic functional molecular crystalline materials has made great progress. A series of special achievements have been made in molecular nanomagnets (including single-molecule magnets and single-chain magnets), spin-transformed materials, MOF magnetic materials, magnetic refrigeration materials, etc., which have made a significant impact on our peers across the world. In particular, in the field of SMMs, we have reported the first case of metallocene rare-earth organic single-ion magnets, which has led the way of this field. In addition, we have proposed a symmetry strategy for constructing rare-earth monomolecular magnets, including metallocene metal–organic structures and pentagonal bipyramidal local coordination structures, which provides new ideas and perspectives for the rational design and controlled

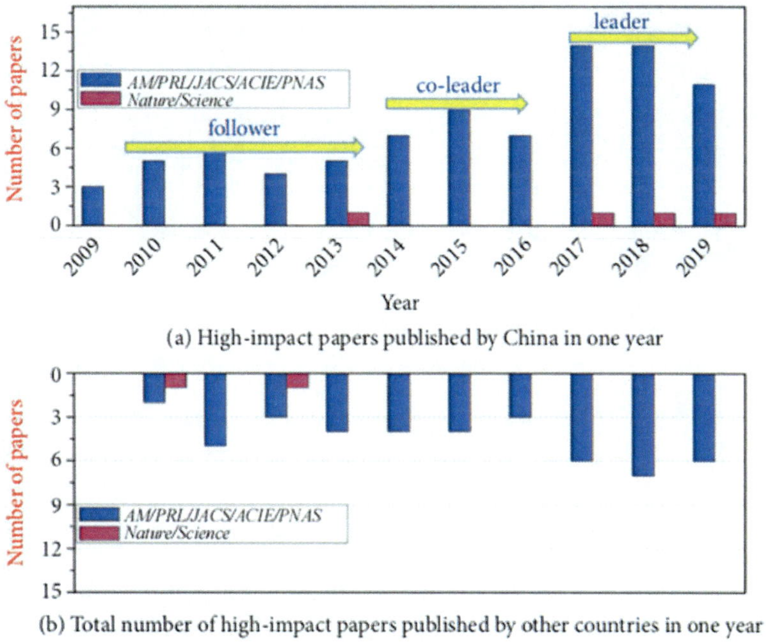

Fig. 2 Leading international research in the field of molecular ferroelectrics

synthesis of high-performance rare-earth monomolecular magnets. The first dysprosium dichloride compound $[Cp^{3+}{}_2Dy]^+$ without equatorial coordination has been constructed using tri-tert-butylcyclopentadiene (Cp^{3+}) with high potential resistance, and its energy barrier (1277 cm^{-1}) and blocking temperature (60 K) have set new world records. Subsequently, a hybrid dichroic sandwich-shaped dysprosium single-molecule magnet was designed and synthesized, whose T_b reached 77 K, the temperature of liquid nitrogen, for the first time.

2.4 Development Trend of Metal–Organic Framework Materials

Currently, the research results of MOF materials in China have generally caught up with those of the leading countries at the international level. The total number of published papers and the total number of high-impact papers in this field rank first and second, respectively (Fig. 3). Some major results, such as the separation and purification of light olefins (especially ethylene and 1,3-butadiene), CO_2 capture, and catalytic conversion of small molecules, have reached internationally leading levels. Therefore, implementing this Plan has greatly promoted the MOF materials research development in China. Presently, domestic and foreign research in the field of MOF materials shows similar development status and trend. On the one hand, excellent

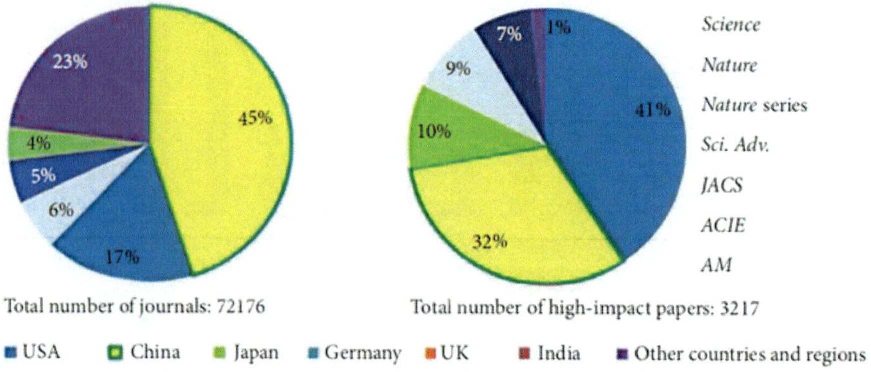

Fig. 3 Percentage of the total number of papers published in the field of MOF materials and the total number of high-impact papers in China

research experts in countries and regions including the USA, Japan, Europe, and China are focusing more on improving MOF materials functions such as adsorption, separation, and catalysis, i.e., high performance; on the other hand, researchers begin to pay more attention to the practicality of high-performance MOF materials, i.e., how to practically apply these materials with the emergence of many high-performance MOF materials these days. Meanwhile, in the past decade, researchers in the field of MOF materials have evolved from the coordination of chemistry scholars at the early stage to scholars in various branches of chemistry. They are now making efforts to become scholars in chemical science, materials science, and even life science. Therefore, cross-disciplinary research with collaboration among scholars from different disciplines has gradually emerged, and the research results have become more diversified and are moving toward practical applications.

2.5 Development Trend of Superconducting Materials

This Plan has achieved two milestone results in the design and controllable preparation of new superconducting function-oriented materials. The first is the discovery of the KFe_2Se_2 series of new iron-selenium-based superconductors, which has opened up a new field for international superconductivity research. More than 350 laboratories in more than 41 countries have followed the work on FeSe-based high-temperature superconductors research, which became one of the most active frontier research in the field of physics from 2012 to 2014. Simultaneously, solid-state ion-grating techniques have been developed to overcome the doping limit of FeSe-based crystalline materials and significantly regulate the carrier concentration of superconducting materials, providing new ideas for exploring new superconducting materials.

Since 2000, the investment in high-temperature superconductive materials (mainly copper oxides) has been greatly reduced in many countries around the world. Still, the continuous support in this direction in China has made a significant development of human resources and software and hardware in superconductivity. Only a few research groups in China have done some highlight studies on copper-oxide high-temperature superconductivity. However, regarding iron-based superconductivity, China's accumulated strength has been fully demonstrated in the study of materials, properties, and mechanisms. *Science*, therefore, published an editorial entitled *"New Superconductors Propel Chinese Physicists to Forefront"*. High-temperature superconducting materials such as KFe_2Se_2, LiOHFeSe, and high-temperature superconductivity at the $FeSe/SrTiO_3$ interface are key research directions pioneered independently by Chinese scientists. According to the research data from Thomson Reuters and other organizations, the "alkali metal intercalated FeSe superconductor" represented by KFe_2Se_2 was rated as the most popular research direction in the whole field of physics in 2013. It was ranked as one of the top ten research topic in the field of physics for two consecutive years from 2013 to 2014. This series of work has dramatically changed the direction of research on non-traditional superconductors and brought the research on FeSe-based superconductors to a new level.

2.6 Trends in Thermoelectric Materials

This Plan has achieved two milestone results in the structural design and controllable preparation of new thermoelectric function-oriented materials. Theoretically, a complete set of defective multi-degree-of-freedom electro-acoustic coupling synergistic control strategies has been proposed, which can significantly improve the thermoelectric properties of materials and motivates thermoelectric materials research; meanwhile, cation cutting techniques have been developed to control the crystal structures of materials, which also significantly improve the thermoelectric properties of materials and provide new ideas for the exploration of new thermoelectric materials. With the support of this Plan, much progress has been made in the field of thermoelectric materials research in China. The existing strategies for optimizing thermoelectricity based on defect engineering have concerned less on the study of other defect-induced modulation degrees of freedom, focusing only on electron-charge and phonon degrees. In addition to the existing electron-charge and phonon degrees of freedom, this Plan has focused on several new types of defect-triggered regulatory degrees of freedom, such as defect-related spin degrees of freedom, defect-based atomic and charge transfer effects, and defect-related surface interface effects, drawing the attention from thermoelectric researchers who are combining them with the existing mainstream optimization strategies to reexamine the optimization of thermoelectric materials based on a "multi-degree-of-freedom synergistic regulation" research idea. This new research concept both expands on the physical concept of defect engineering strategies and injects new life into thermoelectric materials research.

References

1. Fu DW, Cai HL, Liu Y et al (2013) Diisopropylammonium bromide is a High-Temperature molecular ferroelectric crystal [J]. Sci 339:425
2. You YM, Liao WQ, Zhao DW et al (2017) An Organic-Inorganic perovskite ferroelectric with large piezoelectric response [J]. Sci 357:306
3. Ye HY, Tang YY, Li PF et al (2018) Metal-Free Three-Dimensional perovskite ferroelectrics [J]. Sci 361:151
4. Liao WQ, Zhao DW, Tang YY et al (2019) A molecular perovskite solid solution with piezoelectricity stronger than lead zirconate titanate [J]. Sci 363:1206
5. Li PF, Liao WQ, Tang YY et al (2019) Organic enantiomeric High-Tc ferroelectrics [J]. Proc Natl Acad Sci USA 116:5878
6. Zhang HY, Tang YY, Shi PP et al (1928) Toward the targeted design of molecular ferroelectrics: modifying molecular symmetries and homochirality [J]. Acc Chem Res 2019:52
7. Sessoli R, Gatteschi D, Caneschi A et al (1993) Magnetic bistability in a Metal-Ion Cluster [J]. Nature 365:141
8. Ako AM, Hewitt IJ, Mereacre V et al (2006) A ferromagnetically coupled Mn19 aggregate with a record S = 83/2 ground Spin state [J]. Angew Chem Int Ed 45:4926
9. Ishikawa N, Sugita M, Ishikawa T et al (2003) Lanthanide Double-Decker complexes functioning as magnets at the Single-Molecular level [J]. J Am Chem Soc 125:8694
10. AlDamen MA, Clemente-Juan JM, Coronado E et al (2008) Mononuclear lanthanide Single-Molecule magnets based on polyoxometalates [J]. J Am Chem Soc 130:8874
11. Jiang SD (2016) An organometallic Single-Ion magnet [J]. J Am Chem Soc 138:2829
12. Liu JL, ChenY C, Zheng YZ et al (2013) Switching the anisotropy barrier of a Single-Ion magnet by symmetry change from Quasi-D5h to Quasi-Oh [J]. Chem Sci 4:3310
13. Chen YC (2016) Symmetry-Supported magnetic blocking at 20 K in pentagonal bipyramidal Dy(III) Single-Ion magnets [J]. J Am Chem Soc 138:2829
14. Liu J, Chen YC, Liu JL et al (2016) A stable pentagonal bipyramidal Dy (III) Single-Ion magnet with a record magnetization reversal barrier over 1000 K [J]. J Am Chem Soc 138:5441
15. Goodwin CAP, Ortu F, Reta D et al (2017) Molecular magnetic mysteresis at 60 kelvin in dysprosocenium [J]. Nature 548:439
16. Guo FS (2017) A dysprosium metallocene Single-Molecule magnetfunctioning at the axial limit [J]. Angew Chem Int Ed 56:11445
17. Guo FS, Benjamin M, Chen YC et al (2018) Magnetic hysteresis up to 80 kelvin in a dysprosium metallocene Single-Molecule magnet [J]. Sci 362:1400
18. Vincent R, Klyatskaya S, Ruben M et al (2012) Electronic Read-Out of a single nuclear Spin using a molecular Spin transistor [J]. Nature 488:357
19. Godfrin C, Thiele S, Ferhat A et al (2014) Electrically driven nuclear spin resonance in Single-Molecule magnets [J]. Sci 344:1135
20. Kamihara Y, Watanabe T,Hirano M, et al (2008) Iron-Based layered superconductor La [O1-xFx]FeAs (x=0.05−0.12) with Tc = 26 K [J]. J Am Chem Soc, 130: 3296
21. Ren ZA, Lu W, Yang J et al (2008) Superconductivity at 55 K in iron based F-Doped layered quaternary compound Sm[O1-xFx]FeAs [J]. Chin Phys Lett 25:215
22. Guo JG, Jin SF, Wang G et al (2010) Superconductivity in the iron selenide KxFe2Se2 (0<x<1) [J]. Phys Rev B 82:180520
23. Ying TP, Chen XL, Wang G et al (2012) Observation of superconductivity at 30–46 K in AxFe2Se2 (A = Li, Na, Ba, Sr, Ca, Yb, and Eu) [J]. Sci Rep 2:426
24. Ying TP, Chen XL, Wang G et al (2013) Superconducting phases in Potassium-Intercalated Iron selenides [J]. J Am Chem Soc 135:2951–2954
25. Lu XF, Wang NZ, Wu H, et al (2015) Coexistence of superconductivity and antiferromagnetism in (Li0.8Fe0.2)OHFeSe [J]. Nat Mater, 14: 325
26. Ge JF, Liu ZL, Liu C et al (2015) Superconductivity above 100 K in Single-Layer FeSe films on doped SrTiO3 [J]. Nat Mater 14:285

27. Drozdov AP, Eremets MI, Troyan IA et al (2015) Shylin: Conventional superconductivity at 203 kelvin at high pressures in the sulfur hydride system [J]. Nature 525:73–76
28. Drozdov AP, Kong PP, Minkov VS et al (2019) Eremets: Superconductivity at 250 K in lanthanum hydride under high pressures [J]. Nature 569:528–531
29. Biswas K, He J, Blum ID et al (2012) High-Performance bulk thermoelectrics with all scale hierarchical architectures [J]. Nature 489:414–418

Major Research Achievements

Maochun Hong, Yicheng Wu, Chunhua Yan, Yuliang Li, Jiyang Wang, Xianhui Chen, Jingjun Xu, Yongjun Chen, Rong Chen, Baosheng Huang, Xuefeng Fu, Junlin Yang, Shouzhu Zhang, Kexin Chen, Qidong Wang, Jie He, Hongcheng Lu, and Guohong Zou

Since the inception of this Plan, we have been taking full advantage of interdisciplinary cooperation among chemistry, materials, and physics. We have conducted an in-depth study on optical, electrical, and magnetic functions of functional motifs, the fundamental composition and structural basis of crystalline materials that determine their properties, and showed the functional motifs that determine the macroscopic properties of crystalline materials and their spatial integration, providing a theoretical basis for the structural design and controllable preparation of functionally-oriented crystalline materials. This Plan has achieved breakthroughs in several international

M. Hong (✉)
Fujian Institute of Research On the Structure of Matter, Chinese Academy of Sciences, Fuzhou, China
e-mail: hmc@fjirsm.ac.cn

Y. Wu
Tianjin University of Technology, Tianjin, China

C. Yan
Peking University, Beijing, China

Y. Li
Institute of Chemistry, Chinese Academy of Sciences, Beijing, China

J. Wang
Shandong University, Jinan, China

X. Chen
University of Science and Technology of China, Hefei, China

J. Xu
Nankai University, Tianjin, China

Y. Chen · R. Chen · B. Huang · X. Fu · J. Yang · S. Zhang · K. Chen · Q. Wang · J. He · H. Lu · G. Zou
National Natural Science Foundation of China, Beijing, China

© Zhejiang University Press 2023

M. Hong (ed.), *Structural Design and Controllable Preparation of the Function-Directed Crystalline Materials*, Reports of China's Basic Research,
https://doi.org/10.1007/978-981-99-3768-4_3

research frontiers, maintained China's leadership position in the field of crystalline materials, and greatly promoted the integration and intersection of chemistry, materials science, mathematics, information science, and other disciplines. Innovative research results have been achieved in the following five areas.

(i) A new technique for controlling the charge density of single-molecule magnetic functional motifs using a crystal field has been proposed. The functional motifs were used to design and synthesize a series of SMMs, which opened up a new field of single-ion magnets and significantly increased the energy barrier (> 2,200 K). For the first time, they broke the magnetic blocking temperature over the liquid nitrogen temperature range, propelling China's magnetic functional molecular crystalline materials to the forefront of the international market.
(ii) We have greatly improved the ferroelectric properties of molecular ferroelectrics, making them comparable to inorganic ceramic ferroelectrics for the first time. We have developed molecular ferroelectrics with excellent piezoelectric properties that have outperformed commercial multivariate inorganic ceramics. China has advanced from a follower to a leader in the international research frontier of molecular ferroelectrics due to the new area of metal-free chalcogenide ferroelectric materials, which has given molecular ferroelectrics the potential for single-chiral and optical activity.
(iii) We have proposed the atomic response theory of NLO, which has improved the resolution of NLO performance to the atomic level. We were the first in the world to discover several new deep-UV NLO crystal materials, breaking the bottleneck of high NLO effect and high laser damage threshold of infrared NLO crystal materials, and achieving $6 - 11$ μm long-wave infrared laser output, making China the second country after the United States to achieve 10,000 W laser output using laser crystal materials. These innovative research results enable China to continue to dominate the international research frontier in the field of laser crystal material and nonlinear opcital crystal materials.
(iv) The discovery of the new KFe_2Se_2 series of iron-selenium-based superconductors disrupted the superconductivity mechanism of nested electron/hole Fermi surfaces and opened up a new field for international superconductivity research. This is an internationally recognized new superconductor discovered by Chinese scientists and followed by over 350 laboratories in over 30 countries. It maintained the leading position of our iron-based superconductivity research in the world is maintained.
(v) We developed a series of crystalline materials with novel properties, such as MOF materials and 2D crystalline materials, based on molecular designing and crystal engineering principles. We have proposed the academic idea of strengthening molecular recognition by supermolecular interaction in the restricted space, and realized the preferential adsorption of alkanes, reversed adsorption selectivity, and thus improved the purification efficiency of butadiene.

We applied the new method of "phase interface strain" to regulate the lattice strain by two-phase eutectic to produce giant polarized crystalline ferroelectric thin-film materials, increasing the polarization of classical $PbTiO_3$ ferroelectrics by 3.8 times and the highest value of ferroelectrics previously reported internationally by 80% for the first time. We used biomimetic mineralization to produce artificial mother-of-pearl crystalline materials that are highly similar to natural pearl layers. This method can be used to prepare bone grafts with a nacre-like structure and also to prepare a variety of new biomimetic engineering materials.

1 Structural Designing and Modulation of Magnetic Functional Molecular Crystalline Materials

1.1 Ground-State Spin and Uniaxial Magnetic Anisotropy Modulation of Single-Molecule Magnets

Since the discovery of single-molecule magnets (SMMs) in 1993, the goal of high-performance SMMs research has been to improve the effective energy barrier (U_{eff}) and magnetic blocking temperature (T_b). Regulation of the magnetic exchange between metal ions by the structural assembly is an effective way to obtain SMMs with high ground-state spins. The uniaxial magnetic anisotropy of SMMs is mainly regulated by rational control of structural factors such as axial crystal field strength and transverse field symmetry to achieve a breakthrough in the effective energy barrier and blocking temperature of SMMs.

In collaboration with Sato's group at Kyushu University in Japan, Tao Liu's group at the Dalian University of Technology designed and synthesized a highly symmetric ferromagnetically coupled $\{Fe_{42}\}$ high-core cluster. Magnetic tests and theoretical calculations have showed that this complex has a ground-state spin of $S = 45$, which was the highest at that time (Fig. 1).

Song Gao's group at Peking University synthesized and studied β-biketone-based rare-earth single-ion magnets. Their study results have attracted wide attention and

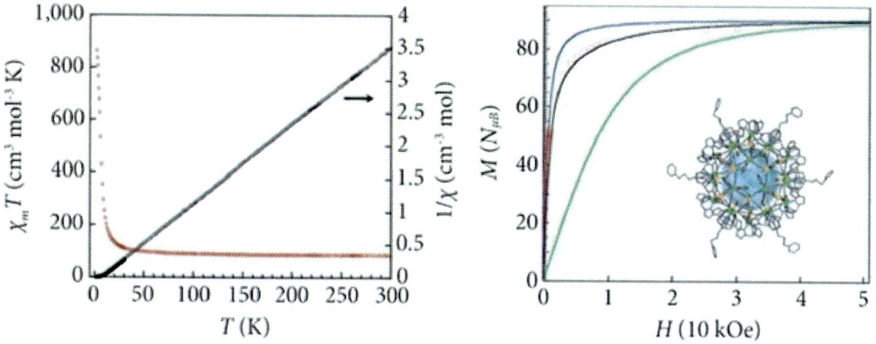

Fig. 1 Highly symmetric ferromagnetic coupled $\{Fe_{42}\}$ hypernuclear cluster with a ground-state spin of $S = 45$

led to related research after its publication, and the study of single-ion magnets soon has become a hot topic in molecular magnetism research. Furthermore, they designed and synthesized a class of Co^{2+} aminophenol complexes with D_3 symmetry using effective Hamiltonian, ligand field, and quantum chemical ab initio methods. The angular momentum of the Co^{2+} ion orbitals in this complex molecule was partially preserved. Its slow magnetic relaxation energy barrier (109 K) at zero external fields was the highest value reported at that time. Song Gao's group systematically investigated the relationship between magnetic anisotropy and properties of rare-earth single-ion magnets using theoretical calculations, and found that the anisotropy of rare-earth Dy^{3+} ions was influenced not only by the symmetry of the geometric structure but also and significantly influenced by the symmetry of charge distribution and ligand bond strength. The group further designed and prepared a high-performance binuclear rare-earth dysprosium single-molecule magnet with a slow magnetic relaxation energy barrier of 721 K, the highest binuclear dysprosium complex at that time. It designed and synthesized rare-earth-phenoxy compounds with nearly linear coordination structures based on the findings. These compounds' Dy^{3+} and Er^{3+} systems have excellent single-molecule magnet properties, among which the relaxation energy barrier of Dy^{3+} single-molecule magnet is as high as 961 K (Fig. 2).

Jinkui Tang's group at the Changchun Institute of Applied Chemistry, Chinese Academy of Sciences elucidated the role of rare-earth single-ion anisotropy and metal-ion interactions in slow magnetic relaxation and proposed an effective strategy for suppressing quantum tunneling effects and raising the energy barrier of SMMs by optimizing the coordination field and improving spin-to-spin magnetic interactions.

Mingliang Tong's group at Sun Yat-sen University was the first to make an innovative theoretical prediction that transverse magnetic anisotropy could be effectively eliminated and magnetic quantum tunneling effect could be suppressed by placing rare-earth ions in C_∞, D_{4d}, D_{5h}, D_{6d}, and other symmetric coordination environments, which laid an important foundation for the appropriate design of rare-earth single-ion magnets with high magnetic blocking temperatures and effective

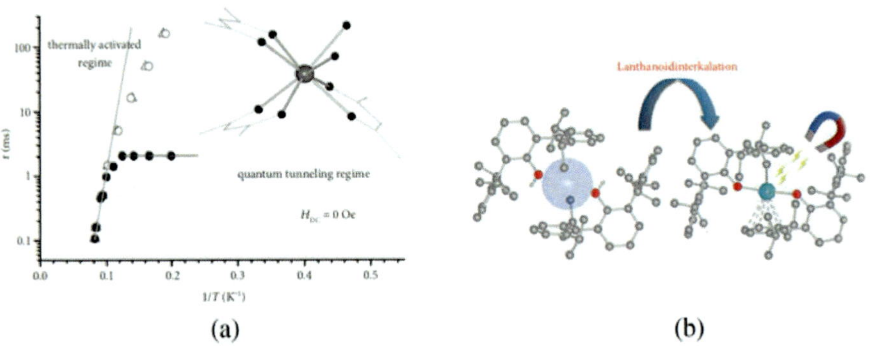

Fig. 2 **a** Rare-earth single-ion magnets based on β-biketones; **b** Rare-earth-enzyme oxygenates with nearly linear coordination structures

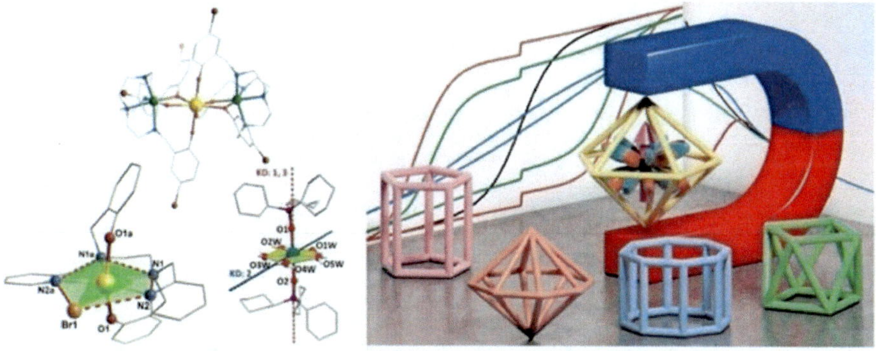

(a) (b)

Fig. 3 **a** Series of pentagonal biconical high-performance rare-earth SMMs; **b** Ligand symmetry-oriented strategy

energy barriers based on crystal field theoretical predictions (Fig. 3). The first high-performance rare-earth single-ion magnet with D_{5h} local symmetry $\{Zn_2Dy\}$ was synthesized, creating a new system of pentagonal biconical rare-earth single-ion magnets. By further strengthening the axial ligand field of the D_{5h} symmetry single-ion magnet, the research group designed and synthesized a Dy^{3+} single-ion magnet with high thermal stability, which set a record for the effective energy barrier of SMMs over 1,000 K. Simultaneously, its thermal stability exceeded 300 °C, laying a solid foundation for the device development of SMMs. The group successfully obtained Dy^{3+} single-ion magnets with strong coordination based on axial phosphine oxide by optimizing the transverse coordination field in the pentagonal plane and continuing to approach the ideal D_{5h} local symmetry, setting a record for blocking temperature up to 20 K. This series of coordination symmetry-oriented rare-earth monomolecular magnet research has greatly contributed to the rapid development of high-performance monomolecular magnets.

Peng Cheng's and Wei Shi's groups at Nankai University used a solvothermal method to synthesize the first isosceles triangular SMMs with toroidal magnetic moments, which have both high-temperature dielectric anisotropy and low-temperature slow magnetic relaxation behavior. Through first-principles calculations, the researchers have showed that the ground-state spins of the three Dy_3 SMMs are arranged in a circular pattern. The AC magnetization exhibits a multi-step slow magnetic relaxation behavior and a characteristic single-crystal hysteresis line, according to low-temperature magnetic properties. This SMM is also a potential ferroelectric one because it crystallizes in the polarity point group. Subsequently, the groups prepared the first six-coordinate dysprosium SMM with a magneto-dielectric coupling effect through solution synthesis. They produced a high-quality millimeter-sized single crystal through coordination between metal ions, ligands, and solvents. They studied its magnetic properties to find *a* slow magnetic relaxation behavior at 40 K. The hysteresis lines in the *a/b* plane showed that the magnetic anisotropy in the *b* direction was stronger than that in the a direction, and the magnetic field

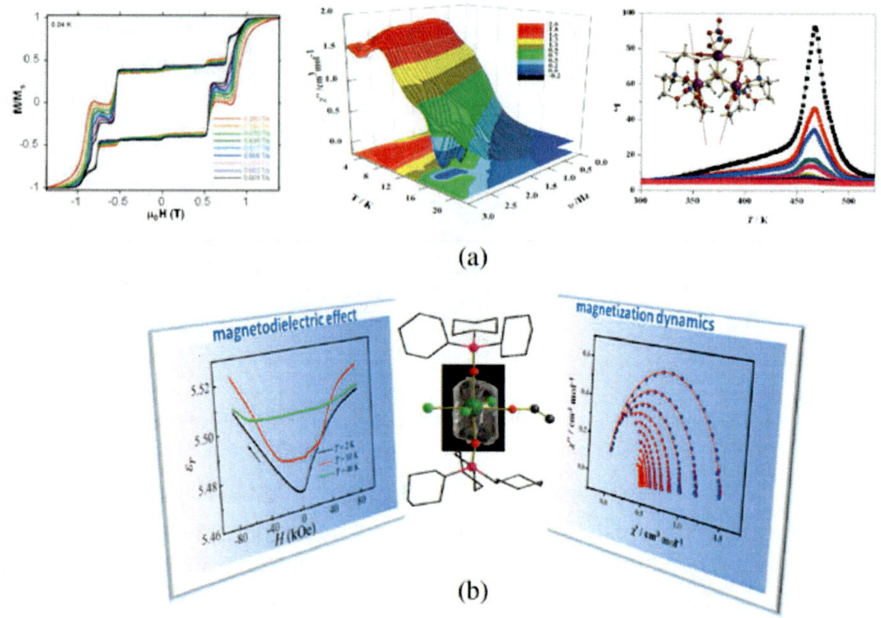

Fig. 4 **a** Toroidal magnetic moments, slow magnetic relaxation behavior, and high-temperature dielectric anomalies of the trinuclear dysprosium monomolecular magnet Dy_3; **b** crystal structure of the six-coordinate dysprosium monomolecular magnet and its monomolecular magnet behavior and magneto-dielectric coupling effect

applied at ±10 T in the *a/b* plane showed that the constant dielectric increases with the increase of the magnetic field, showing a significant magnetic dielectric effect. Furthermore, the hysteresis lines in the *b* direction are stronger than those in the a direction, which is consistent with the hysteresis lines of single crystals. The theoretical analysis suggests that the magnetic dielectric coupling effect originates from the strong spin–lattice coupling of trivalent Dy^{3+} ions in the high spin state, which results in the change of molecular capacitance (Fig. 4). This has revealed that the intrinsic magneto-dielectric coupling effect could also be realized in non-long-range magnetically ordered SMMs. Eventually, the electric field can regulate the magnetic properties of SMMs, which opens up a new direction for the research on SMMs and magneto-dielectric coupling materials.

1.2 Metal–Organic Single-Ion Magnets

Metal–organic single-ion magnets have flexible coordination numbers, abundant bonding modes, and rare electronic structures compared with classical coordination compounds, which provide a broad research scope for SMMs. The first metal–organic single-ionic magnet cyclopentadienylcyclooctatetraenyl erbium with a slow relaxation energy barrier of 323 K, was reported by Song Gao's research group in 2011. This work introduced rare-earth metal–organic chemistry to the study of high-performance SMMs and attracted more inorganic, organic and coordination chemists to the field of SMMs.

Mingliang Tong's research group designed and characterized quasi-linearly coordinated dysprosium monomolecular magnets in collaboration with Layfield's research group, setting new world records for the effective energy barrier (>2,200 K) and blocking temperature (80 K) of monomolecular magnets, with blocking temperature exceeding the liquid nitrogen temperature (77 K) for the first time, laying the foundation for monomolecular magnets' practical application.

The magnetic properties of transition-metal ions, unlike rare-earth metal ions, are mostly derived from *d* electrons, whose orbital the ligand field easily quenches angular momentum, and the electronic structure and magnetic energy levels are highly influenced by the ligand field. The magnetic relaxation energy barrier of transition-metal SMMs is generally small (mostly < 100 K). Therefore, Co^{2+} ions, whose orbital angular momentum is difficult to be quenched, are an effective magnetic source for the construction of transition-metal monomolecular magnets. Song Gao's group has studied an example of a Co = N monomolecular magnet with quasi-linear coordination mode and super strong magnetic coupling, which has the highest relaxation energy barrier of 578 K for transition-metal monomolecular magnets at that time, based on their understanding of the relationship between transition-metal coordination field symmetry and magnetic anisotropy. It is worth noting that cobalt's orbital angular momentum is fully preserved; the cobalt magnetic behavior is similar to that of rare-earth ions, which opens up a new way and a new idea to regulate the magnetic properties of transition-metal ions by chemical means. They were invited to write a review paper on "Influence factors and regulation of magnetic anisotropy in single-ion magnets."

Due to zero-field splitting and spin–orbit coupling, Cr^{2+}, Mn^{3+}, and Fe^{2+} have d^4, d^4, and d^6 electronic configurations, respectively, and can generate strong magnetic anisotropy under appropriate ligand fields, and are therefore promising candidates for SMMs. Cr^{2+} is the most reactive of the three, and only a few cases of corresponding SMMs have been reported. Semi-sandwich-type metal carbonyl complexes [e.g., $ArCr(CO)_3$ and $CpCo(CO)_2$] are simple molecular models that have been widely studied in the field of metal–organic chemistry. The loss of one electron in these complexes results in highly reactive 17-electron metal–organic radicals, which have important applications in chemical reactions and catalytic transfer. Still, the use of semi-sandwich-type carbonyl compounds to prepare magnetic molecular materials has rarely been reported. The research group of Xinping Wang and You Song at Nanjing University developed and synthesized weakly coordinated anion-stabilized semi-sandwich metal carbonyl binuclear complexes $Cr(CO)_3(\eta^6,\eta^5\text{-}C_6H_5C_5H_4)Co(CO)_2$ and $[Cr(CO)_3]_2(\eta^6,\eta^6\text{-}C_6H_5C_6H_5)$. The first heteronuclear metal–metal half-bond radical with near-infrared absorption (1,031 nm) was discovered in the study of oxidation of the former, which has potential applications in optoelectronic communication; however, single or double oxidation of $[Cr(CO)_3]_2(\eta^6,\eta^6\text{-}C_6H_5C_6H_5)$ yielded the same decomposition, which underwent single-crystal to single-crystal transformation in pentane to Cr^{2+} complexes with single-molecule magnet properties, making it one of the few examples of chromium-centered SMMs so far.

1.3 Assembly Based on High-Performance Single-Ion Magnet Substrates

The controlled assembly of high-performance SMMs is aided by single-ion magnets with large magnetic anisotropy as building blocks and inorganic or organic ligands as bridging ligands. The group of Xinyi Wang at Nanjing University tried to develope a single-molecule magnet with a high T_b through anisotropic magnetic exchange between $[Mo(CN)_7]^{4-}$ and Mn^{2+} ions. It successfully produced a single-molecule magnet with a $\{Mn_2Mo\}$ trinuclear structure. It had the highest energy barrier among the cyano-bridged SMMs at the time, and very large hysteresis lines and quantum tunneling steps could be observed at low temperatures. The group also synthesized another example of $\{Mn_2Mo\}$ trinuclear monomolecular magnet with a higher energy barrier based on the same strategy, which can be transformed from single-crystal-to-single crystal to extranuclear $\{Mn_4Mo_2\}$ compound by dynamically opening and forming coordination bonds through water loss, enabling dynamic ground-state spin transition, regulated monomolecular magnet, and anisotropic magnetic exchange in the single-crystal state. Furthermore, they have dynamically regulated the color transformation of the complex in a single-nucleated cobalt complex through a dynamic reversible water gain/loss process and also induced a reversible transformation in the behavior of the spin-crossover and SMMs. This is the first complex to achieve reversible transformation of spin-crossover and single-molecule magnet behavior through single-crystal to single-crystal transformation.

Metal–organic frameworks are three-dimensional ordered structures formed by metal-ion nodes and organic linkers connected by coordination bonds, characterized by structural diversity and a wealth of chemical and physical properties. The single-crystal-to-single-crystal conversion of Peng Cheng's group and Wei Shi's group increased the effective energy barrier of slow magnetic relaxation of Dy_2 SMMs as nodes from nearly zero to more than 100 K, allowing for reversible exchange of guest molecules between the rare-earth MOF systems $[Dy_2(INO)_4(NO_3)_2]\cdot 2DMF$ and $[Dy_2(INO)_4(NO_3)_2]\cdot 2CH_3CN$. By ab initio calculations, it was found that this change in the effective energy barrier was not due to an endogenous energy level difference between the Dy_2 nodes but to a two-order-of-magnitude difference in their quantum tunneling rates. They also synthesized a trinuclear dysprosium single-molecule magnet crystallized in the polar space group. This compound has electric hysteresis lines at room temperature and magnetic hysteresis lines at low temperatures and is an example of a magnetic molecular aggregate with both electromagnetic quadratic stability.

Jinkui Tang's group successfully prepared {Dy_6} single-molecule magnet composed of two {Dy_3} units "edge-to-edge." To optimize the ring magnetic moment, it modulated the ring magnetic moment unit connection and intramolecular magnetic interactions to enable the assembly and isotropic enhancement of two ring magnetic moment functional motifs. In addition, they presented a holistic approach to the design of rare-earth SMMs from the perspective of structure and ligand design.

Mingliang Tong's group further replaced the antimagnetic ion Zn^{2+} with paramagnetic Fe^{2+} ions to set a new record for the effective energy barrier of 3d – 4f heterometallic monomolecular magnets, and the presence of iron allowed the magnetic blocking phenomenon to be observed in the Musburger spectrum based on the high-performance rare-earth monotonic magnet {Zn_2Dy}.

The Yanzhen Zheng research group at Xi'an Jiaotong University obtained the first rare-earth-based "nanotube" {Dy_{72}} with both slow magnetic relaxation and high plasmonic conductivity. The outer wall of this tubular molecule consists of 72 rare-earth metal centers connected by hydroxyl ions. It is a hollow circular tube with an inner diameter of about 0.7 nm and a length of 2.8 nm. The inner and outer surfaces of the tube wall composed of metal hydroxyl compounds were highly hydrophilic, and the water between the guest ions inside the pores formed a superionic conductor (proton conductivity of 1.2×10^{-2} S/cm) in a high-temperature and high-humidity environment through a hydrogen-bonding network. Simultaneously, the molecule has slow relaxation behavior and is the largest rare-earth single-molecule magnet with the largest nucleus number.

Tao Liu's group used $[Tp^{*}Fe(CN)_3]^{-}$ to build unit-bridging Fe^{2+} spin-transformation functional motifs and form high spin one-dimensional chains through photoinduced Fe^{2+} spin-transformation, resulting in the successful construction of a photoinduced single-chain magnet. They realized the reversible charge transfer in the {Fe_2Co} trinuclear cluster, self-assembled the charge-transferring motif {Fe_2Co} into a one-dimensional magnetic chain, matched the redox potential of metal ions by modulating the octahedral coordination field, and reversibly induced the charge transfer between Fe and Co using laser lights of different wavelengths. They discovered a new approach to writing and erasing information at the molecular level by enabling and inhibiting single-chain magnets' delayed magnetic relaxation behavior. They wrote a review paper to explain the reflections and progress of using light to regulate the electronic structure and multifunctional coupling of molecular materials based on the studies of photoinduced spin-transformation-driven multifunctional synergy.

1.4 Nuclear Spin Strategy for Suppressing Magnetic Quantum Tunneling Effect

The magnetic quantum tunneling effect is very important in SMMs, and nuclear spins can affect the properties of SMMs through hyperfine interactions. However, the hyperfine interactions are very weak. It is difficult to observe the nuclear spins acting on the magnetic relaxation process and the magnetic quantum tunneling effect in SMMs. A new strategy to suppress the zero-field magnetic quantum tunneling effect of single-ion magnets by nuclear spins was proposed by Mingliang Tong's group. For the first time, hyperfine interactions were observed in $^{165}Ho^{3+}$ single-ion magnets with pentagonal biconical coordination geometry to effectively suppress the magnetic quantum tunneling effect at zero magnetic fields, providing a new strategy for improving the performance of SMMs with nuclear spin suppression of magnetic quantum tunneling effect.

2 Design, Synthesis, and Performance of High-Performance Molecular Ferroelectrics

Ferroelectric crystals are a class of functional crystalline materials with outstanding electrical, force, optical, acoustic, thermal, and magnetic properties that respond to external effects such as electric field, stress, pressure, temperature, and light using electrodes. They are flexible, lightweight, easy to process, low-cost, and environment friendly, and are expected to be a beneficial complement to commercial inorganic ferroelectrics. Focusing on the overall objectives of the Plan, Chinese scholars developed new methods for designing molecular ferroelectrics via molecular modifications and introduction of chirality, such as "quasi-spherical theory" "introducing homochirality" and "hydrogen/fluorine substitution". These methods achieve the chemical designing of molecular ferroelectrics without relying on crystal databases or blind searching.

Funded by this Plan, researchers in China have made several breakthroughs in the field of molecular ferroelectrics. Rengen Xiong's research group discovered in 2013 that the Curie temperature (T_c) of diisopropylamium bromide (DIPAB) reached 426 K and the saturation polarization (P_s) reached 23 $\mu C/cm^2$, making the first time that the T_c and P_s of this molecular ferroelectric achieved the level of $BaTiO_3$ (Barium titanate). Subsequently, in 2017, the same group designed and synthesized molecular ferroelectric materials TMCM-$MnCl_3$ and TMCM-$CdCl_3$ with a piezoelectric coefficient (d_{33}) of 185 pC/N and 220 pC/N, respectively. These values exceeded those of all previous molecular materials while approaching $BaTiO_3$ levels. Then, in 2018, they fabricated the first metal-free organic perovskite ferroelectric, adding a new member to the important family of perovskite materials. Recently, the group discovered the first molecular ferroelectric solid solution with an MPB and soft molecular ferroelectrics with piezoelectric properties comparable to those of hard commercial inorganic ferroelectric ceramics. This group was also the first to discover the organic enantiomeric ferroelectrics. These innovative research results have showed that Chinese scholars were progressing from followers to neck-to-neck runners and eventually to leaders in the international molecular ferroelectrics industry.

2.1 *Molecular Ferroelectrics with High the Curie Temperature, Large Saturation Polarization*

Molecule-based ferroelectrics are flexible, lightweight, low-cost, easy to process in solution, and biocompatible, which can eliminate the problems of high hardness, high density of heavy metals, difficult processing, and high-energy consumption in the current widely used inorganic oxide ferroelectrics. However, T_c of molecular ferroelectrics is mostly below room temperature, and P_s is relatively low, which is not helpful for their practical applications. How to improve the ferroelectric properties of molecular ferroelectrics is a crucial and challenging issue in the molecular ferroelectrics research.

Rengen Xiong's research group found that T_c of diisopropylamine bromide (DIPAB) reached 426 K, and its P_s was as high as 23 $\mu C/cm^2$. This was the first time since the discovery of the first ferroelectric resonate in 1920 that T_c and P_s of a molecular ferroelectric were comparable to those of the inorganic oxide ferroelectric barium titanate an important breakthrough in ferroelectric molecular research. *Chemical & Engineering News* commented as follows: "They found that DIPAB had spontaneous polarization similar to that of barium titanate as well as several other benchmark ferroelectric features. Furthermore, DIPAB can be processed from an aqueous solution, making it easy to use and sustainable." "DIPAB has a saturation polarization similar to barium titanate, a greater dielectric constant than ferroelectric polymers, and a coercive field one-hundredth of polymers and half of barium titanate, resulting in excellent energy savings," according to a guest comment from *Science*. "DIPAB, with its advantages of easy processing and continuous environmental friendliness, may be able to replace oxide ferroelectrics in various applications due to its ferroelectric phase stability and outstanding properties. Furthermore, this organic compound has outstanding ferroelectric behavior, significant piezoelectricity, and electrostriction effects. These three properties have recently been shown in local observations of biological or physiological material compositions. Therefore, the ferroelectric coupling is likely to be an important part of some biological processes. In this case, the multifunctional DIPAB may be a bridge between oxide and composite soft material coupling."

The group has also designed and synthesized a large class of high T_c metal-free organic chalcogenide ferroelectrics [$A(NH_4)X_3$, where A is a divalent organic cation and X is a halogen ion], among which MDABCO-NH_4I_3 has a high T_c of 448 K and a large P_s of 22 $\mu C/cm^2$ (Fig. 5) and is comparable to barium titanate.

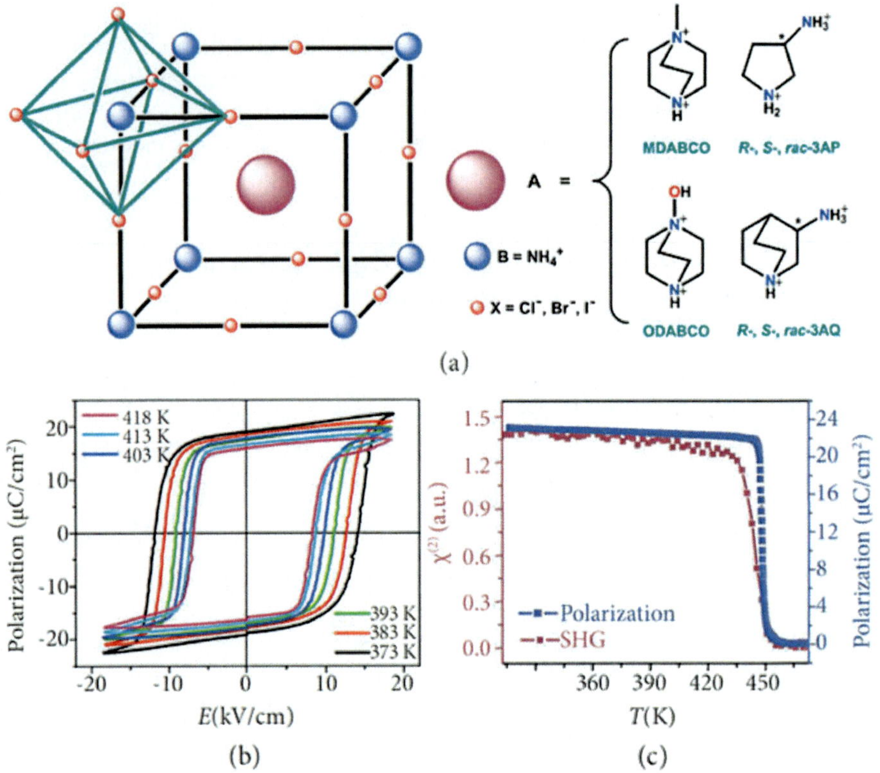

Fig. 5 **a** Structure of three-dimensional metal-free chalcogenide ferroelectrics; **b** Hysteresis lines of MDABCO-NH_4I_3; **c** Polarization intensity of CMDABCO-NH_4I_3 and the second-order NLO response intensity as a function of temperature

Chalcogenides are a class of functional materials that includes inorganic chalcogenides and organic–inorganic hybrid chalcogenides, which have been one of the focuses in physics, chemistry, and materials science research due to their excellent ferroelectric, piezoelectric, optoelectronic and catalytic properties. There had never been any reports of metal-free organic chalcogenide ferrites before; therefore, this effort contributed a new class of functional chalcogenide materials to the family. Since no metal elements are required, metal-free chalcogenides can eliminate metal toxicity and high preparation cost to meet practical materials' energy-saving, economic, and environmental protection requirements. Because of the significance and leadership of this research, its findings were named among China's top ten scientific and technological advancements in 2018. Meanwhile, *Science* published a review titled "Perovskite Ferroelectrics Go Metal-Free," in the same issue, pointing out that metal-free organic chalcogenide ferrites were finally comparable to $BaTiO_3$. Stoddart, Editor-in-Chief of *Nature Reviews Materials*, cited this project as a research highlight and introduced it under the topic "Purely Organic Perovskites," and metal-free organic chalcogenide ferrites are were expected to be applied in chiral catalysis and optical switching, in addition to organic systems.

2.2 *Molecular Ferroelectrics with Large Piezoelectric Coefficients*

The direct conversion of mechanical and electrical energy is accessible to the piezoelectricity of ferroelectrics, which is one of the most essential and widely used features. Although molecular ferroelectrics are approaching or surpassing inorganic ferroelectrics in performance parameters such as the Curie temperature and saturation polarization strength, their piezoelectric properties have always been their shortcoming (piezoelectric coefficient $d_{33} < 50$ pC/N). There has never been a molecular piezoelectric with its d_{33} comparable to $BaTiO_3$ (190 pC/N) or PZT (200 – 750 pC/N) since the discovery of the piezoelectric effect in 1880.

Renege Xiong's research group designed and synthesized the molecular ferroelectrics $[(CH_3)_3NCH_2Cl]MnCl_3$($TMCM-MnCl_3$) and $[(CH_3)_3NCH_2Cl]CdCl_3$ ($TMCM-CdCl_3$) with d_{33} as high as 185 pC/N and 220 pC/N, respectively. These d_{33} values exceeded all previous molecular materials and were comparable to the inorganic ferroelectric $BaTiO_3$. This result solved the century-old problem of insufficient piezoelectricity of molecular materials and provided new ideas and directions for researching molecular piezoelectric materials.

Although the d_{33} of the single-component $TMCM-CdCl_3$ (220 pC/N) has already surpassed that of the single-component $BaTiO_3$ (190 pC/N), it is still far lower than those of the mainstream multi-component ferroelectric ceramic solid solutions, such as the solid binary solution PZT (200 – 750 pC/N) and the multi-component solid-solution $Pb(Mg_{1/3}Nb_{2/3})O_3$-$0.3PbTiO_3$ (1,500 pC/N). Further increasing the d_{33} of molecular ferroelectrics is a huge difficulty. In 1954, Jaffe et al. were the first to discover the existence of MPBs in PZT, and various physical properties, especially piezoelectric properties, were maximized at the MPBs. However, no solid molecular solution with MPB has been found for more than 60 years. Rengen Xiong's group found for the first time that the molecular chalcogenide $[(CH_3)_3NCH_2F]CdCl_3$ ($TMFM-CdCl_3$) based on TMFM could be used to construct molecular chalcogenide solid solution $(TMFM)_x$ $(TMCM)_{1-x}CdCl_3$ $(0 \le x \le 1)$ with $TMCM-CdCl_3$ based on the similar structural parameters of $[(CH_3)_3NCH_2F]^+$(TMFM) and TMCM (Fig. 6). The room-temperature phase changed from ferroelectric monoclinic phase m-point group $(0 \le x < 0.25)$ to ferroelectric phase hexagonal 6 mm-point group $(0.3 < x \le 0.5)$, then to non-ferroelectric phase hexagonal 6/m-point group $(0.55 \le x \le 1)$, and the high-temperature phase was always the hexagonal phase 6/*mm*-point group, as TMFM content increases. The morphotropic phase boundary was clearly observed in the composition range of $0.25 \le x \le 0.3$, where the ferroelectric monoclinic phase m and the hexagonal phase 6 mm coexist, unlike the MPB in PZT, where the ferroelectric tripartite and tetragonal phases coexist, which was the first observation of MPB in a solid molecular solution. The piezoelectric coefficient significantly increased at MPB, and d_{33} reached seven times that of the single-component TMCM-$CdCl_3$ (220 pC/N). This discovery can be called a disruptive technology in the field of piezoelectricity, enabling flexible molecular ferroelectric materials to be comparable to hard commercial inorganic ferroelectric ceramics in terms of piezoelectric properties, opening up a wide space for the science and application of molecular piezoelectric materials, and greatly promoting the development of molecular ferroelectrics.

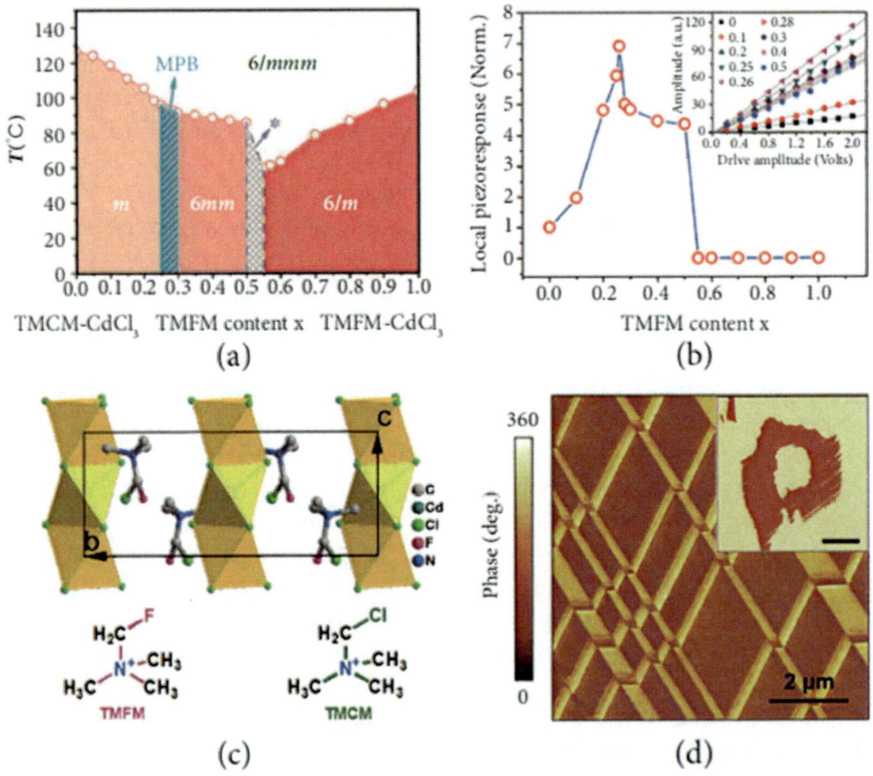

Fig. 6 **a** Phases of $(TMFM)_x(TMCM)_{1-x}CdCl_3$ $(0 \leq x \leq 1)$; **b** Relative piezoelectric response of $(TMFM)_x(TMCM)_{1-x}CdCl_3$ $(0 \leq x \leq 1)$ at different components; **c** Structure of $(TMFM)_x(TMCM)_{1-x}CdCl_3$ $(0 \leq x < 0.25)$ and schematic representation of the structures of TMFM and TMCM; **d** Ferroelectric domains and polarization reversal of domains of $(TMFM)_x(TMCM)_{1-x}CdCl_3$ $(x = 0.25)$

2.3 Homochiral Molecular Ferroelectrics

In 1920, a homochiral molecular molecule was found as the first example of a ferroelectric rosinate. The rosinate, as the first chiral ferroelectric, highlighted the possibility of merging chirality and ferroelectricity to explore a broader variety of applications. Ferroelectricity and chirality, which are both related to spatial symmetry breaking, have a closer intrinsic symmetry relationship than ferromagnetic, superconducting, and negative thermal expansion features. Five of the 10 polar point groups that allow for ferroelectricity are chiral, namely point groups 1 (C_1), 2 (C_2), 4 (C_4), 3 (C_3), and 6 (C_6). However, the study of chiral ferroelectrics has long been neglected due to the absence of chiral centers in the widely studied inorganic ferroelectric ceramic materials.

Rengen Xiong's group has found the unusual homochiral multi-axis molecular ferroelectric (*R*)-3 hydroxyquinic ring halide, and the spin-switching effect generated by the changing of the optical axis has potential optoelectronic applications. Chiral ferroelectrics have largely been multi-component organoamine salts or metal complexes for the past century. The first pair of single-molecule organic enantiomeric ferroelectrics, (*R*)-3-quinuclidinol and (*S*)-3-quinuclidinol, and the racemic (*R*)-3-quinuclidinol, were reported by the group of Rengen Xiong. The chiral (*R*)-3-quinuclidinol and (*S*)-3-quinuclidinol crystallize at room temperatures in the enantiomeric polar group C_6, and the vibrational circular dichroism spectra and crystal structures exhibit a perfect mirror relationship (Fig. 7). Both enantiomers exhibit 622*F*6-type ferroelectric phase transitions with Curie temperatures up to 400 K, comparable to the classical inorganic ferroelectric barium titanate (393 K). Furthermore, the saturation polarization intensity (up to 7 $\mu C/cm^2$) is comparable to that of the organic polymer ferroelectric PVDF (polyvinylidene fluoride), and the low coercive field (15 kV/cm) ensures easy reversal of ferroelectric polarization, which has a promising application prospect for memory devices and optoelectronic devices. However, their racemic (*R*)-3-quinuclidinol is not ferroelectric. This research reveals the great advantage of homochirality in the precise designing of high Curie point ferroelectrics and provides an effective way to further explore high-performance chiral organic ferroelectrics.

2.4 Molecular Ferroelectrics with Semiconductor Properties

Ferroelectric semiconductors with both ferroelectricity and semiconductor properties are an important class of optoelectronic functional materials. Ferroelectric semiconductors have a unique anomalous photogenerated voltammetric effect, where the photogenerated voltage can be much higher than the forbidden band width of the material. They are promising for a wide range of applications in photovoltaic devices, optical drivers, and optical sensors due to their high photovoltaic voltage and

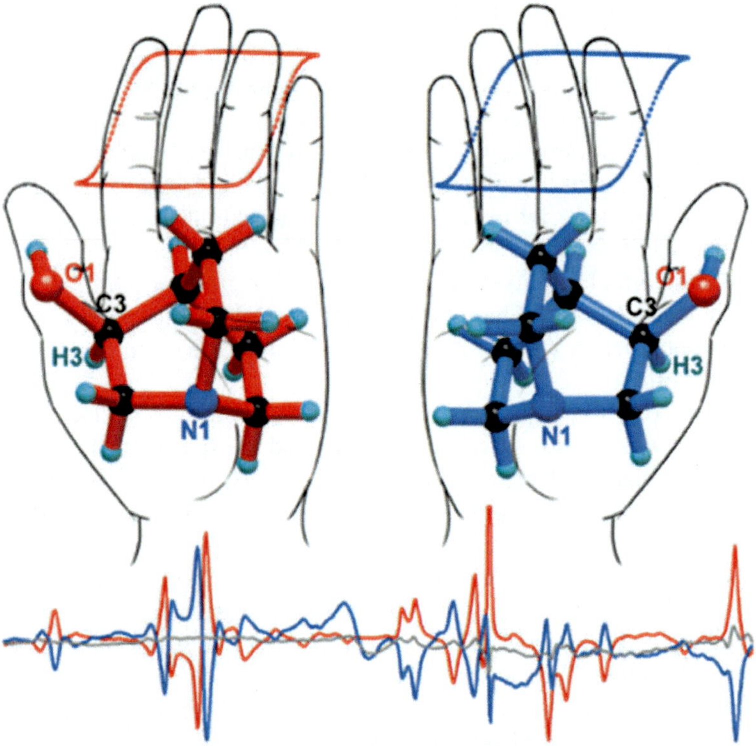

Fig. 7 Molecular structures of (*R*)-3-quinuclidinol and (*S*)-3-quinuclidinol and their vibrational circular dichroism spectra exhibit perfect mirror image relationship

customizable photovoltaic properties of the electric field. The research on ferroelectric semiconductors has mainly focused on inorganic ferroelectrics, such as $BaTiO_3$ and $BiFeO_3$. Organic monomolecules, organic amines, crown ether inclusions, and metal-formic acid frameworks are examples of molecular ferroelectrics that do not exhibit semiconductor properties.

Rengen Xiong's research group designed and synthesized the first molecular ferroelectric semiconductor (benzylammonium)$_2$ $PbCl_4$ with an organic–inorganic hybrid 2D chalcogenide structure, a Curie temperature of 438 K, a spontaneous polarization intensity of 13 $\mu C/cm^2$, and a direct semiconductor band gap of 3.65 eV. Ferroelectric semiconductors had never been reported before, and this research lit up the path to constructing molecular ferroelectric semiconductors. Subsequently, the group designed and synthesized a 2D chalcogenide molecular ferroelectric semiconductor $[CHA]_2[PbBr_4]$ (CHA is a cyclohexylamine cation) with a lower band gap and regulated its ferroelectric and semiconducting properties by replacing a portion of Br with I. The band gap (2.74 eV) of the molecular ferroelectric semiconductor $[CHA]_2[PbBr_{4-4x}I_{4x}]$ ($x = 0.175$) is comparable to that of the widely studied inorganic ferroelectric semiconductor $BiFeO_3$ (2.7 eV). Junhua Luo's research group found that $[CHA]_2[PbBr_4]$ crystals exhibited anisotropic semiconductor photoelectric characteristics under illumination conditions. The 2D extension of the metal skeleton layer produced highly temperature-dependent photovoltaic voltage and photovoltaic current, while the vertical direction exhibited significant photoconductivity properties.

Based on three-dimensional lead bromide chalcogenide, Junhua Luo's group constructed a ferroelectric semiconductor $(C_4H_9NH_3)_2(CH_3NH_3)_2Pb_3Br_{10}$ with a 2D multilayer chalcogenide structure, featured with a spontaneous polarization intensity of 2.9 $\mu C/cm^2$ and an optical band gap of 2.4 eV respectvely, by introducing mixed organic cationic ligands.

The photodetector assembled with this ferroelectric crystal has high detection performance, especially its response time reaching about 150 μs, and it could achieve high sensitivity and fast detection of light radiation in the intrinsic absorption region of the crystal. The group subsequently designed and synthesized a 2D bilayer chalcogenide ferroelectric semiconductor $(C_4H_9NH_3)_2CsPb_2Br_7$ containing organic cations and inorganic alkali metals, introducing flexible organic cations into the three-dimensional $CsPbBr_3$ chalcogenide, with a spontaneous polarization intensity of 4.2 $\mu C/cm^2$ and an optical band gap of 2.70 eV. It was found that, unlike other calixarene-type molecular ferroelectrics, the spontaneous ferroelectric polarization of this compound was induced by the atomic displacement of inorganic alkali metal Cs^+ and the ordered disorder of organic cations. The photoelectric devices assembled on the basis of this ferroelectric crystal also have high detection performance. These works expanded the research of molecular ferroelectrics in optoelectronic applications.

2.5 New Strategies for Precise Design of Molecular Ferroelectrics

Molecular ferroelectrics combine the merits of flexibility, lightweight, and low acoustic impedance found in molecular materials with the excellent properties of ferroelectric materials such as ferroelectricity, piezoelectricity, high dielectricity, and pyroelectricity, which have important academic value and great potential for practical applications. However, the search for new molecular ferroelectrics is like looking for a needle in a haystack and faces a tremendous challenge because ferroelectrics must crystallize in the ferroelectric phase at 10 polar point groups 1 (C_1), 2 (C_2), m (C_s), 2 mm (C_{2v}), 3 (C_3), 3 m (C_{3v}), 4 (C_4), 4 mm (C_{4v}), 6 (C_6), 6 mm (C_{6v}), and often needs a ferroelectric phase transition. Since the discovery of the first ferroelectric, the molecular ferroelectric rosinate, in 1920, the search for ferroelectrics has relied on blind screening. There has been lack of a valide method for targeted synthesis.

From the chemical perspective, Rengen Xiong's research group proposed a new method for the precise design of molecular ferroelectrics by deeply understanding the 10 polar point groups of ferroelectric phases and combining the Curie symmetry principle, the Neumann principle, and the Landau phase transition theory. The new approaches, such as "quasi-spherical theory," "introducing homochirality," and "hydrogen/fluorine substitution" transformed the discovery of molecular ferroelectrics from a blind search to a rational chemical design. Specifically, the "quasi-spherical theory" is a chemical design idea for crystal symmetry reduction, i.e., designing and regulating ferroelectricity by chemically modifying or clipping high-symmetry cations to change the symmetry and specific interactions of crystals at the molecular level. Five out of the 11 chiral point groups are polar point groups (C_1, C_2, C_4, C_3, and C_6). The introduction of chiral molecules made it easier for the material to crystallize in the five chiral polar point groups, which greatly increased the possibility of inducing ferroelectricity, namely the "introducing homochirality" strategy. In addition, the "hydrogen/fluorine substitution" strategy is similar to the "hydrogen/deuterium isotope effect" where the introduction of fluorine atoms usually leads to a slight structural disruption while the polar groups remain unchanged, thus significantly increasing the Curie temperature and spontaneous polarization intensity of the material. Based on these design strategies, the group has precisely synthesized a series of molecular ferroelectrics with excellent properties, such as high saturation polarization high piezoelectric coefficient, semiconducting properties, multipolar axis properties, and photoluminescence.

The chemical design strategy proposed by Rengen Xiong's group has been applied and validated by domestic and foreign peers as a proven method for designing molecular ferroelectrics. In view of this, the group was invited to write a perspective article entitled "Molecular design principles for ferroelectrics: Ferroelectrochemistry" for the *Journal of the American Chemical Society* and introduced the concept of "ferroelectrochemistry" (Fig. 8) to provide an effective methodological guide for the development of high-performance molecular ferroelectrics.

2.6 *Photoferroelectric Semiconductor Crystal Materials for Next-Generation Photoelectric Detection Technology*

Photoferroelectrics are a class of ferroelectric materials with excellent optoelectronic properties due to the mutual coupling of photogenerated carriers and ferroelectric polarization. They have important application prospects for next-generation optoelectronic devices. Photoferroelectrics are a special kind of polar optoelectronic materials in which the internal dipoles are arranged orderly to form spontaneous polarization and can be reversed or reoriented under an applied electric field. Based on the spontaneous polarization, ferroelectric materials exhibit their rich physical and physiological properties, especially under light irradiation, and produce novel photoferroelectric phenomena, such as anomalous photovoltaic effect, photorefractorization, and photomorphogenic effect. The main goal of Junhua Luo's research group was to create new photoferroelectric crystal materials with strong photoelectric coupling and to induce spontaneous polarization by regulating the interaction

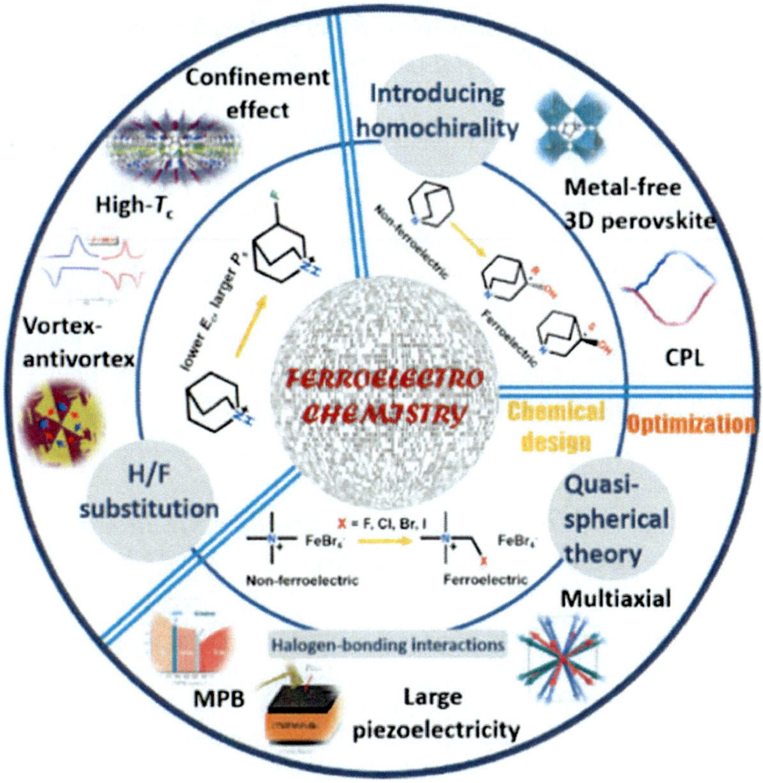

Fig. 8 Schematic of the concept of "ferroelectrochemistry"

between dipole elements. On this basis, they successfully combined ferroelectric polarization with carrier transporting and opened up a new field of research on hybrid "photoferroelectric semiconductors." They developed next-generation photoelectric detection technology driven by iron electrodes, highly sensitive polarization detection technology, and high-performance, high-energy ray detection technology. Relevant research results were published in top international journals such as *Angewandte Chemie International Edition*, *Journal of the American Chemical Society*, and *Advanced Matorials*, leading the development of photoferroelectric crystal materials in the field of optoelectronics. These research projects not only provide a new design strategy for the subsequent designing of photoferroelectric semiconductor crystals, but also lay a foundation for the application of these materials in the optoelectronic industry.

2.7 Discovery of Molecule-Based Ferroelectric Crystalline Materials with Fast Polarization Reversals

How to achieve fast reversals of spontaneous polarization is an important issue that needs to be addressed in the current research on molecular ferroelectric materials. Junhua Luo's group has designed and synthesized an ionic ferroelectric compound and achieved fast spontaneous polarization reversals. It was found that the *N*-methylmorpholine cation and the trinitrophenol anion were connected by a strong N–H...O hydrogen bond, both of which undergo an order–disorder transition at a temperature of 315 K, which induces spontaneous polarization of the compound. The spontaneous polarization of the material can be reversed easily by an applied electric field, with a record reversal frequency of 263 kHz.

2.8 Discovery of Incommensurate Structure Modulated Molecular Ferroelectric Crystals

The incommensurate phase is a special modulated structure of ferroelectric materials, which is important for revealing the origin of ferroelectricity and elucidating the material's physical properties. Compared with metal oxides, there are few studies on the incommensurate phase-modulated structure of molecule-based ferroelectrics, making it very difficult to understand the cause of ferroelectric materials and develop new ferroelectric material systems. During the structural transition of the paraelectric phase to the ferroelectric phase, Junhua Luo's group discovered three consecutive transition state incommensurate phases for the first time. The X-ray diffraction spectra show obvious satellite scattering spots, indicating that the atoms are modulated along the polar axis. The group analyzed the superstructure of the incommensurate phase and found that the period of the modulated structure of the crystal was seven times the period of the primary cell, and the modulated wave vector q along the polar axis was about 0.1589, which was a significant non-integer multiple. In particular, the incommensurate phase modulation structure and the ferroelectric properties of the material were closely related to the disordered structural motifs. In the paraelectric phase, the anionic structural motifs of the compound were highly disordered; in the incommensurate phase, the disorder of the motifs was partially frozen, and some atoms were in the modulated state; in the ferroelectric phase, the atomic motion was further restricted, and the structural motifs were completely ordered, thus inducing a symmetry breaking to produce ferroelectric properties. This successive change of structural motifs leads to a structural phase transition of the crystal, which goes through a paraelectric phaseincommensurate phaseferroelectric phase in sequence. The results of this study described the incommensurate phase structure of molecule-based ferroelectrics in detail, revealed the correlation between the ferroelectricity of the crystal and the structural phase transition, and elucidated the structural causes of the ferroelectric properties of the material, which provided an important reference for the development of new ferroelectric material systems.

3 Laser Crystals and Nonlinear Optical Crystals

Laser crystals and NLO crystals are the material basis for laser generation and laser wavelength expansion. Therefore, the research on new laser crystal and NLO crystal materials has been one of the highlights in the field of inorganic, nonmetallic materials in the past 50 years. The current commercial laser crystals and NLO crystals frequency generation and conversion range have covered a wide range of optical wavelengths from mid-infrared to ultraviolet. In the field of NLO crystals, China's crystal research power, marked by the made-in-China crystals, has formed a completely innovative technology chain from theoretical design and material preparation to product domination in the global market, leading international research on new crystals. As the demand for laser applications expands to the deep ultraviolet and mid-and far-infrared wavelengths, China aims to originality innovations on the present basis and strive to maintain its international leadership. In the field of laser crystals, although China has made important contributions to the growth of some crystals such as titanium gemstones and yttrium vanadate, the practical laser crystals were innovated by other countries which are also controlling some important laser crystals. As the laser evolves toward high power, wide tuning, and high frequency, the existing commercial crystals can no longer meet the application requirements in some aspects. Our goal is to develop new laser crystals independently, expand the applied bands, and make breakthroughs in "stranglehold" technologies. Using the functional-motif method, we constructed optoelectronic functional motifs, studied the relationship between the structure and performance of optoelectronic functional motifs and the effective synergy of optoelectronic functions, developed inorganic modular chemical synthesis methods based on optoelectronic functional motifs, systematically explored the structure–property relationship theory for NLO crystals and laser crystals, NLO crystals in deep-UV and mid-infrared wavelengths, and special-wavelength laser crystals. The research aimed to achieve the structural design and controlled synthesis of optoelectronic functional substances. Our research on UV NLO crystals continues to take the lead worldwide. Some of the research on IR crystals have entered the international frontier, and several "stranglehold" technological challenges have been overcome due to breakthroughs in various applications of crystalline materials.

3.1 Research on the Structure–Property Relationships of Nonlinear Optical Crystals

The anion group theory proposed by Academician Chuangtian Chen in 1976 has played a key role in synthesizing and designing boronate UV NLO crystals, which has promoted leapfrog development in this field. However, with the development of laser technology, the demand for new crystals has become particularly urgent as the requirements for short-wave UV NLO crystals are increasing. After years of research on UV crystalline borate systems from simple borates to complex borates, especially beryllium borates, which have been studied more systematically in structure design and crystal growth, it is increasingly difficult to find new borate NLO crystal materials. Therefore, it is imperative to develop new UV materials based on the theory of anionic groups.

Starting from the core functional motif of borates, namely triangular [BO_3] groups, Ning Ye's group at the Fujian Institute of Research on the Structure, for the first time proposed using the CO_3 groups, which also have a π-conjugated electronic structure, as a functional motif to extend the exploration of NLO crystal materials to carbonate systems. The coplanar parallel arrangement of CO_3 groups was achieved and a series of new carbonate UV NLO crystals with the molecular formula $ABCO_3F$ (A = Na, K, Rb, Cs; B = Ca, Sr, Ba, Zn, Cd, Pb) were produced by precisely regulating the relative sizes of alkali metals and divalent alkaline earth, d_{10} transition metals Zn and Cd, and lone pair electron-containing cations such as Pb in the lattice (Fig. 9). Similar to borates, these carbonates characterized by the parallel arrangement of CO_3 groups, feature short UV absorption edges, high birefringence, high NLO coefficients, and high laser damage thresholds, many of which can achieve phase matching in the short-wave UV band (266 nm, quadruple frequency). Related research has attracted wide attention and interest from scholars at home and abroad. Professor Halasyamani from the University of Houston synthesized $KMgCO_3F$, $RbMgCO_3F$, and $Cs_9Mg_6(CO_3)_8F_5$ and grew $KSrCO_3F$ crystals to about 4 cm, which experimentally verified their excellent performance for short-wave UV frequency doubling. Ning Ye's group further investigated the structure–property relationships of these compounds. Their results revealed the microscopic assembly of CO_3 groups to cation radius ratio in these compounds, which further uncovered the regulation of cation radius ratio and NLO coefficient. In addition, the group found that the unique p-π interaction between Pb^{2+} and CO_3 groups in $CsPbCO_3F$ crystals leads to an anomalous and strong second-order multiplication effect (Fig. 10).

In addition, Ning Ye's group used other novel strategies to explore new carbonate NLO crystal materials.

(i) When applying the molecular engineering approach to the design and synthesis of carbonate crystals, $YCa_4O(BO_3)_3$ (YCOB for short) and $KBe_2(BO_3)F_2$ (KBBF for short) were used as template compounds to synthesize two carbonate NLO crystals with potential applications in short- and even deep-UV bands $Ca_2Na_3(CO_3)_3F$ (Fig. 11) and $NaZnCO_3(OH)$ (Fig. 12).

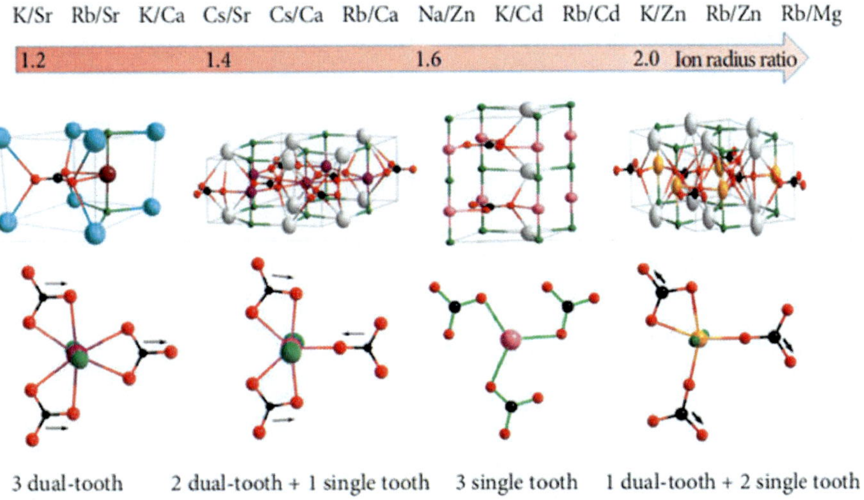

Fig. 9 Structural characteristics of $ABCO_3F$-like crystals and the regulation of anionic groups by cations

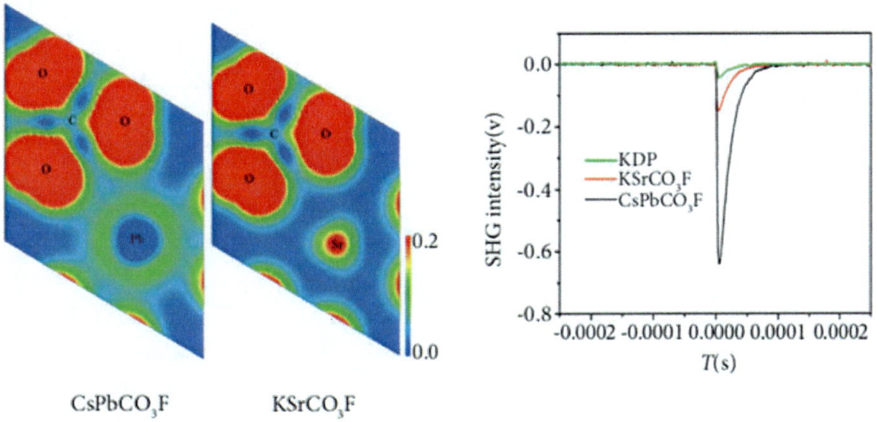

Fig. 10 p-π effect in $CsPbCO_3F$ crystals

(ii) $K_3VO(O_2)_2CO_3$ (KVCO for short) crystals were synthesized by introducing the d_0 cation (V) polyhedra and the fixed-domain $[O_2]^{2-}$ peroxy group into the carbonate system. The great enhancement of the multiplication effect by the fixed-domain π orbital group was discovered for the first time, providing a new idea for designing crystals with high NLO coefficients (Fig. 13). The isomorphic $Rb_3VO(O_2)_2CO_3$ and $Cs_3VO(O_2)_2CO_3$ crystals synthesized by Kang's research group at Chung-Ang University in Korea and Guohong Zou's group at Sichuan University respectively exhibited super-frequency-multiplication effects.

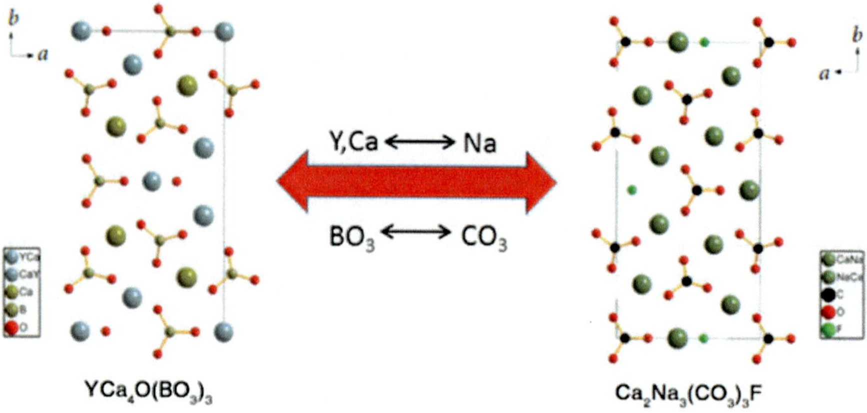

Fig. 11 Comparison of the crystal structures of $Ca_2Na_3(CO_3)_3F$ and $YCa_4O(BO_3)_3$

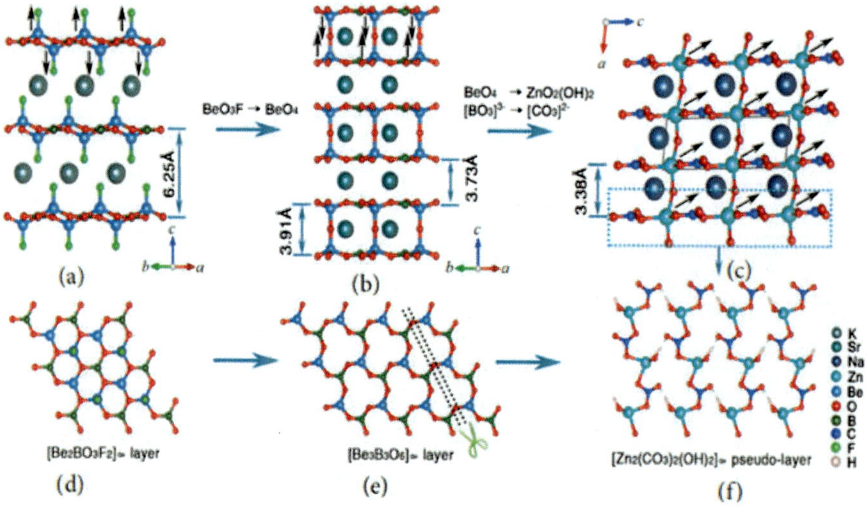

Fig. 12 Comparison of the crystal structure of $NaZnCO_3(OH)$ with those of KBBF and SBBO

(iii) For the first time, a partial substitution strategy of aliovalent cations was applied to the design and synthesis of deep-UV carbonate NLO crystal materials, where the centrosymmetric $Y_2(CO_3)_3 \cdot H_2O$ (YC for short) was transformed into the noncentrosymmetric $(NH_4)_2Ca_2Y_4(CO_3)_9 \cdot H_2O$ (CYC for short) while the original framework structure was maintained, thus giving the crystal a multiplication effect of 2.1 KDP and a short UV cutoff (shorter than 200 nm) (Fig. 14).

(iv) For the first time, four isomorphic hydroxyl halide carbonates with different NLO coefficients, $Re_8O(CO_3)_3(OH)_{15}X$ (Re = Y, Lu; X = Cl, Br), were synthesized by introducing halogen ions (Cl, Br) with high electronegativity

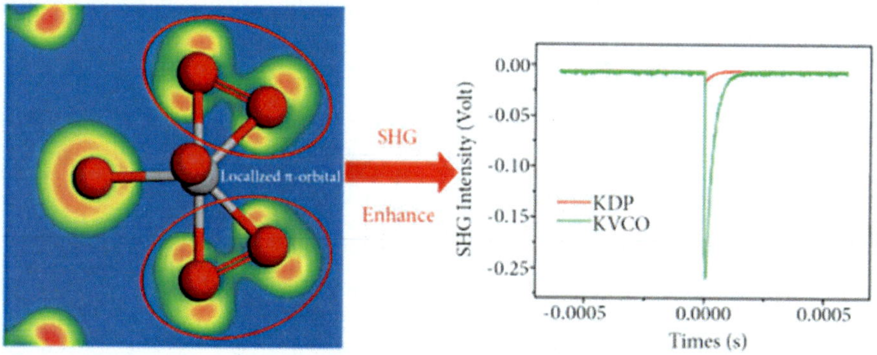

Fig. 13 Mechanism of frequency enhancement by fixed-domain π-orbit $[O_2]^{2-}$ peroxy groups

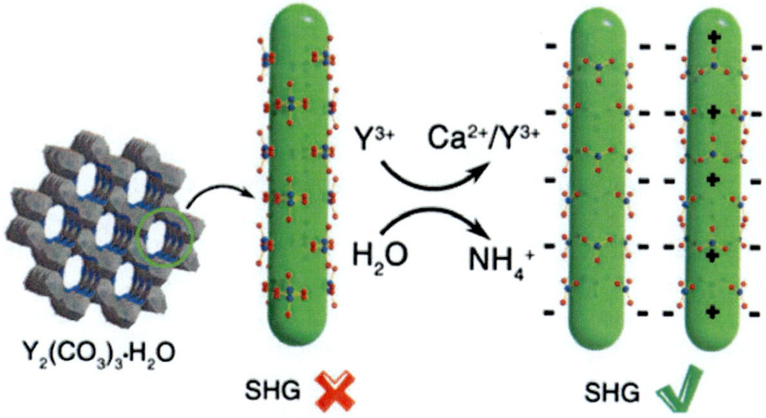

Fig. 14 Partial substitution strategy of aliovalent cations for the transformation of centrosymmetric YC to noncentrosymmetric CYC crystals

as equilibrium anions into the rare-earth carbonate system. It was found that the differences in their octave coefficients mainly resulted from the differences in the local field correction factors (F), which are related to their refractive indices. Thus, the octave coefficients of this series of compounds are modulated by their different refractive indices. Based on a similar π-conjugated group design, Ning Ye's research group also explored several NLO nitrate crystals in nitrate systems with non-aqueous solubility and significant effects, including several Pb-containing nitrate compounds with non-centric structures, a series of non-aqueous rare-earth hydroxyl nitrate crystals $Re(OH)_2NO_3$ (Re = La, Y, Gd), as well as the first NLO fluoronitrate crystal $Pb_2(NO_3)_2(H_2O)F_2$ with a super-multiplication effect (12 KDP).

Ning Ye's group further extended the design idea of triangular π-conjugated group to the organic field. In the guanidinium salt system, using the classical $KBe_2BO_3F_2$

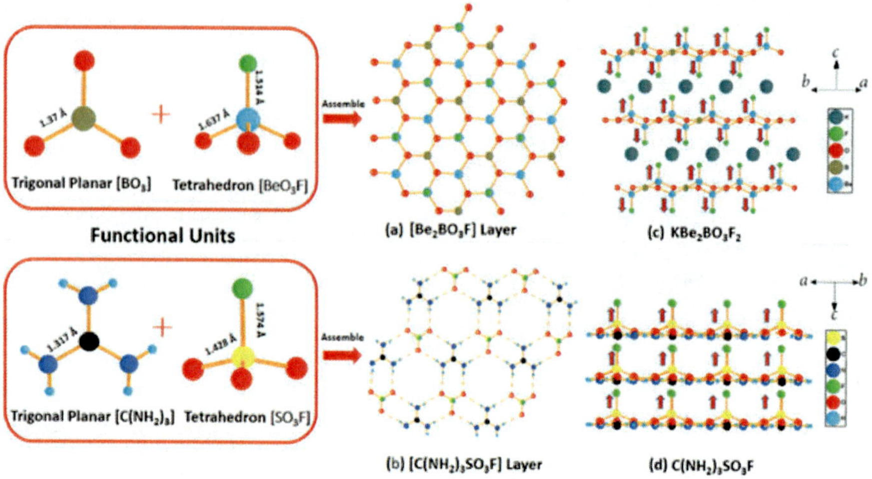

Fig. 15 Structural design of $C(NH_2)_3SO_3F$ crystal

(KBBF) structure as a template, the BO_3 group and the tetrahedral $(BeO_3F)^{5-}$ group in KBBF were replaced by the organic planar triangular $[C(NH_2)_3]^+$ group and the inorganic tetrahedral $(SO_3F)^-$group, respectively. They successfully obtained the first fluorosulfonate UV NLO crystal, i.e., $C(NH_2)_3SO_3F$ (Fig. 15). The crystal is metal-free and benefits from the parallel coplanar arrangement of planar trigonal $[C(NH_2)_3]^+$ groups with a high multiplication effect (5 KDP) and a suitable birefringence (0.133@1,064 nm). In addition, calculations showed that the dispersion curves of the crystal were relatively smooth, which enabled phase matching over the transmission wavelength range (the shortest phase-matching wavelength is about 200 nm).

Furthermore, considering that the π-conjugated planar anion group $(B_3O_6)^{3-}$ in BBO can produce a significant effect, from the perspective of the structure–property relationship, Ning Ye's research group proposed using the hydrogen cyanurate $(HC_3N_3O_3)^{2-}$ group, which also has a large π-conjugated electronic structure of a planar six-membered ring, as the functional motif. A series of alkali metal cyanurates $AB(HC_3N_3O_3)\cdot 2H_2O$ (A = K, Rb; B = Li, Na) with NLO effects were successfully synthesized in a green and efficient way by introducing the drop milling method from organic eutectic synthesis and combining with powder multiplication test in the field of nonlinear optics (Fig. 16), and several centimeters of $KLiHC_3N_3O_3\cdot 2H_2O$ were grown in aqueous solutions. The optical properties of these two crystals were systematically characterized, and it was confirmed that both could achieve phase matching below 266 nm and reach a laser damage threshold of about 5 GW/cm^2, making them next-generation UV NLO crystals with promising application prospects (Fig. 17).

Shilie Pan's research group at the Xinjiang Institute of Physical and Chemical Technology, Chinese Academy of Sciences, proposed that the introduction of F to tetrahedral $(PO_4)^{3-}$ groups results in significant microscopic polarizability

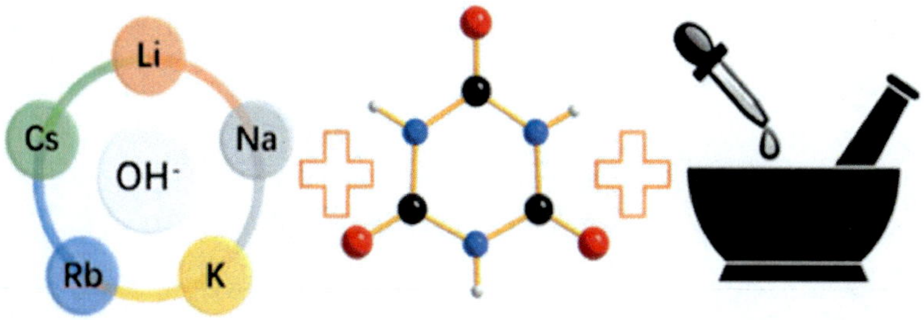

Fig. 16 Exploration of $AB(HC_3N_3O_3)\cdot 2H_2O$ (AB = KLi, RbLi, RbNa, CsNa) crystals Exploring new ionic organic NLO crystals by combining titrimetric milling and neutralization reaction

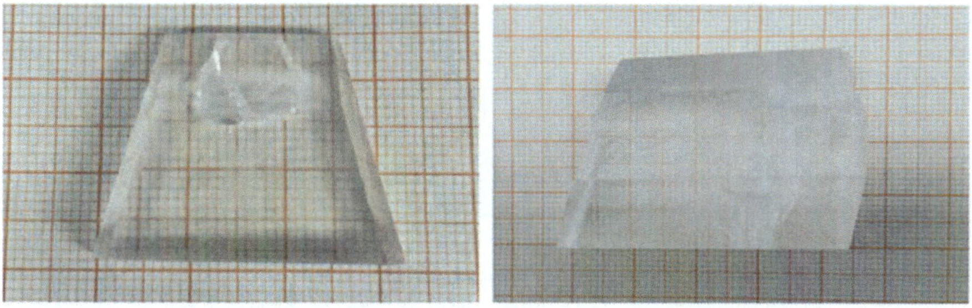

(a) KLHCY (b) RLHCY

Fig. 17 Photo of crystals

anisotropy and microscopic hyperpolarizability for $(PO_3F)^{2-}$ and $(PO_2F_2)^{1-}$ microscopic motifs, which is beneficial for the enhancement of birefringence and frequency doubling effects. The group searched the inorganic crystal structure database, selected $(NH_4)_2PO_3F$ using the first-principles approach, and grew centimeter-scale single crystals using solvent evaporation. The powder doubling experiments showed that the phase matching of 266 nm frequency-doubled light could be satisfied. The fluorophosphate proposed in this work has expanded the research area in this field and verified the effectiveness of design strategy. The research group of Ling Chen at Beijing Normal University discovered a new deep-UV NLO active group $(PO_3F)^{2-}$ with a high optical band gap and second-order superlattice. The group further discovered a series of deep-UV monofluorophosphates and successfully grew $NaNH_4PO_3F\cdot H_2O$ large-sized single crystals using the solution method (Fig. 18). Junhua Luo's group at the Fujian Institute of Research on the Structure, Chinese Academy of Sciences, proposed using non-π-conjugated anion group $(SO_4)^{2-}$ to construct new deep-UV NLO crystals and, on this basis, discovered two deep-UV sulfate crystals, $NH_4NaLi_2(SO_4)_2$ and $(NH_4)_2Na_3Li_9(SO_4)_7$. These studies have greatly expanded the types of anionic groups and led to the discovery of many new UV NLO crystals.

The key to the success of the anion group theory in guiding the design and synthesis of UV crystals is the strong ionicity of the A-site metal cations (usually alkali or

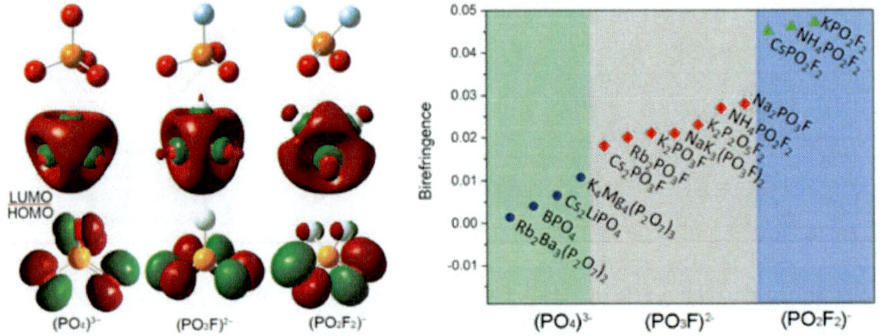

Fig. 18 Calculated and tested birefringence values of the front orbitals of the microscopic groups $(PO_4)^{3-}$, $(PO_3F)^{2-}$ and $(PO_2F_2)^{-}$, typical phosphates, monofluorophosphates, and difluorophosphates, which have proved that the introduction of $(PO_3F)^{2-}$ and $(PO_2F_2)^{-}$ facilitates birefringence

alkaline earth metals) in these crystals, which have a spherical distribution of the peripheral electron cloud and a weak response to the second-order perturbation of the external photoelectric field. It can be neglected in the first-order approximation. The electron cloud distortion caused by the strong covalent bonds in the anion group is the main contributor to the NLO effects. When NLO crystals are explored in the infrared band, the introduction of transition metals or heavy metal cations of the IIIA – VA group is often required to enhance the frequency multiplication effect. In this case, a strong covalent polyhedral structure is formed between the metal and surrounding ions. The electron clouds between the ions overlap significantly, making it difficult to distinguish between the individual cation and anion groups, thus limiting the applicability of the anion group theory. However, the current commercial infrared NLO crystals, such as $AgGaS_2$, $ZnGeP_2$, and $AgGaSe_2$, generally have a low laser damage resistance threshold. Introducing alkali and alkaline earth metals to widen the material's band gap effectively improves the crystal's damage threshold. A systematically theoretical study of infrared NLO crystals of halogen, sulfur, and oxides by Zheshuai Lin's research group at the Institute of Physical and Chemical Technology, Chinese Academy of Sciences, showed that the frequency multiplication effect of infrared NLO crystals mainly comes from the anion group when the A-site cations are alkali and alkaline earth metals, which is consistent with the anion group theory. Therefore, the anion group theory can be extended to the field of mid-infrared nonlinear optics, which can still serve as guidelines for the continuous exploration of new high-quality infrared NLO crystals.

The use of cascaded second-order nonlinear frequency conversion (twice multiplication or once multiplication and once summation) is a common method for obtaining UV-band lasers. However, the second-order nonlinear process can only be generated in noncentrosymmetric crystals, such as β-BBO, LBO, CLBO, CBO, and KBBF, which can meet the practical requirements. In contrast, the third-order nonlinear process is not limited by the crystal symmetry and can be generated in

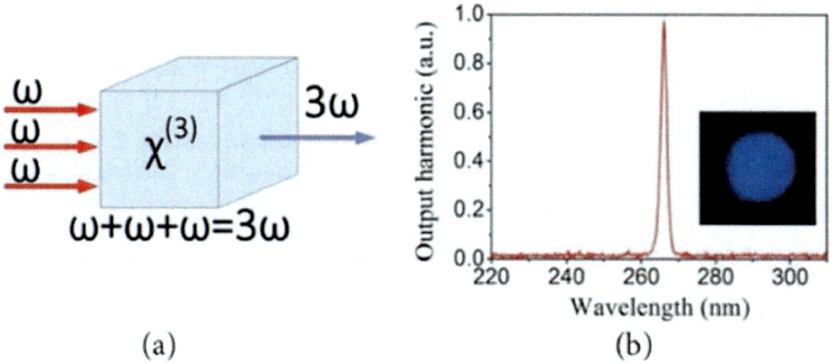

Fig. 19 **a** Third-order nonlinear frequency conversion; **b** 266 nm third harmonic output intensity

centrosymmetric crystals. However, the nonlinear effects in centrosymmetric crystals have been neglected for a long time due to low third-order nonlinear polarizabilities of most crystals. Ning Ye's group achieved the first effective UV laser output in a centrosymmetric crystal by applying third-order nonlinear frequency conversion, and the highest energy of the 266 nm third harmonic was 37.6 μJ with a maximum conversion of 2.5% by choosing β-BBO and calcite crystals with an off-domain π-conjugated structure. The feasibility of using third-order nonlinear frequency conversion to obtain UV lasers and the possibility of obtaining deep-UV lasers in centrosymmetric crystals were proved, showing the potential of third-order nonlinear frequency conversion for practical applications and shining a light on new research directions for the exploration of UV and deep-UV nonlinear crystals (Fig. 19).

With the rapid development of nonlinear optics, China must understand the electronic and atomic-scale origins of NLO crystal materials from a new perspective in the face of new materials, systems, and problems. Based on a general analysis of the electronic structure origin and symmetry constraints of NLO effects, Shuiquan Deng's group at the Fujian Institute of Research on the Structure, Chinese Academy of Sciences, proposed a universal partial response functional method and developed an atomic-scale atomic response theory for nonlinear optics. This theory not only explains the success of the anion group theory, but also provides a quantitative analysis of all the ionic contributions to the crystal under a unified standard, thus raising the understanding of the origin of NLO effects from the anion group theory to the atomic response theory. Based on this theory, the group explored the origin of the frequency multiplication effect of the deep-UV phosphate $LiCsPO_4$ and a series of borate NLO compounds. It revealed the cations' important contribution to the atomic-scale's multiplication effect. By studying chalcopyrite-based infrared NLO crystal materials, they found a simplified relationship between NLO response, polarizability, and band gap. They proposed a simple and practical semi-empirical pattern for designing NLO materials. Based on an advanced genetic evolutionary algorithm for crystal structure prediction, the group further achieved efficient theoretical predictions in sulfur compounds and obtained some new infrared

NLO compounds in collaboration with labs, including Ga_2Se_3 with a SHG coefficient 2.3 times that of $AgGaS_2$, $Na_2ZnSn_2Se_6$ and $Na_2CdSn_2Se_6$ with three times and 2.2 times that of $AgGaS_2$, respectively (Fig. 20). The NLO atomic response theory reveals the polarizable characteristics of the orbitals and the important role of the long-neglected non-occupied orbitals on the multiplication effect. It explains the NLO multiplication effect from a new perspective, which is of great scientific significance for understanding the micromechanism of NLO crystal materials and function-oriented material design.

The groups classified by chemical stability are widely used in the material genome project and the functional group theory of NLO crystal materials. Theoretically,

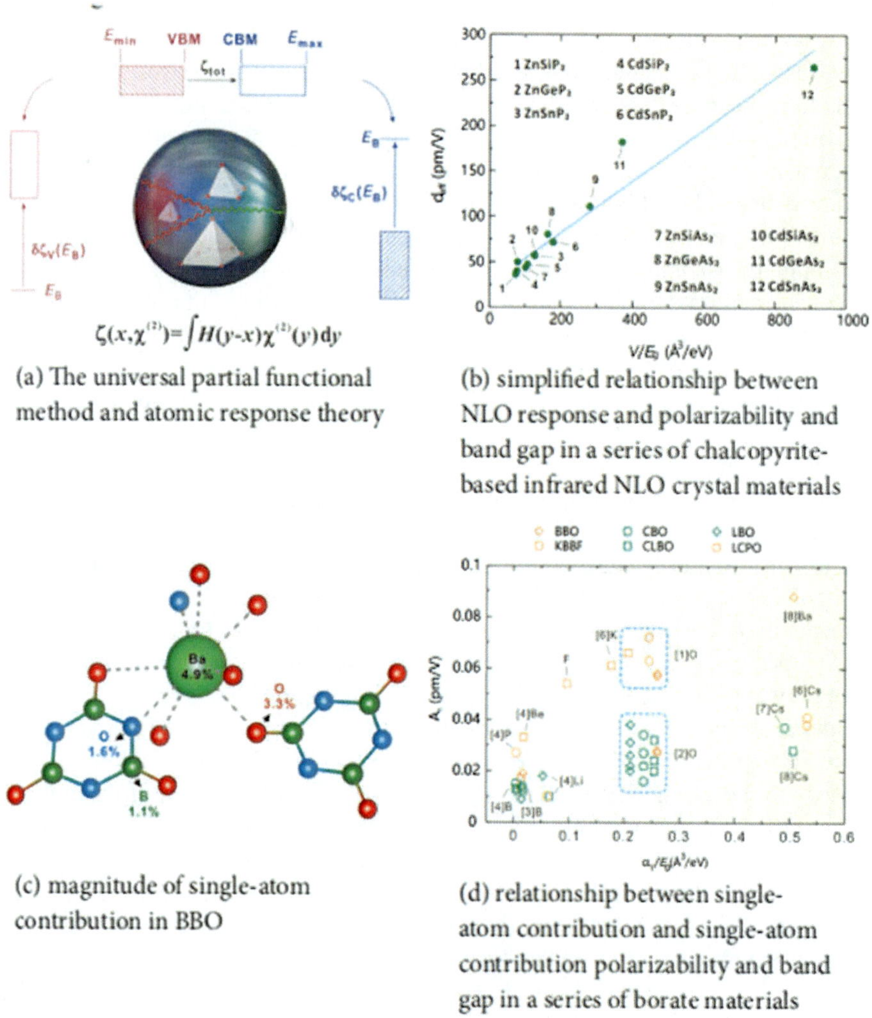

(a) The universal partial functional method and atomic response theory

(b) simplified relationship between NLO response and polarizability and band gap in a series of chalcopyrite-based infrared NLO crystal materials

(c) magnitude of single-atom contribution in BBO

(d) relationship between single-atom contribution and single-atom contribution polarizability and band gap in a series of borate materials

Fig. 20 Principle and application of NLO atomic response theory

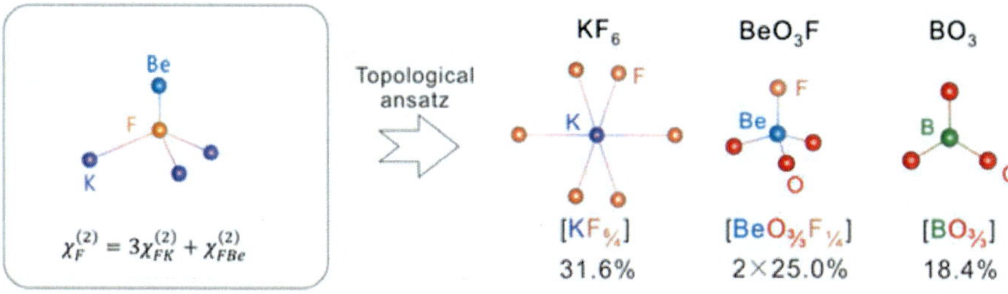

Fig. 21 The group classification scheme and the contribution of each group to the frequency multiplication factor d_{11} in KBBF

the group classification, however, is not necessarily the same for different physical properties. It is hard to compare various combinations of motifs and accurately obtain the functional motifs of materials due to the lack of a structural classification scheme according to physical properties. By analyzing the neutron diffraction differential cross-section and X-ray scattering phenomena, Shuiquan Deng's group at the Fujian Institute of Research on the Structure, Chinese Academy of Sciences, for the first time proposed a general theory of group classification based on physical properties and the contribution of boundary atoms based on a topological scheme (Fig. 21). This theory can be used to calculate the contribution of the groups based on specific physical properties and provides an effective quantitative criterion for the classification of functional groups. The research group illustrated the application of the theory using KBBF as an example.

The Ga_2S_3 microcrystals reported earlier by Guocong Guo's group at the Fujian Institute of Research on the Structure, Chinese Academy of Sciences, have optical properties such as a high laser damage threshold of ten times that of $AgGaS_2$, but their multiplication factors are only comparable to that of commercial $AgGaS_2$. The second-order octave response can be improved by replacing S with more polarizable Se, according to the theory of NLO atomic response. Prof. Shengping Guo of Yangzhou University discovered that the multiplication factors of Ga_2S_3 powder was significantly improved, which was 2.3 times higher than that of $AgGaS_2$, and was characterized by a high laser damage threshold, a wide optical transmission range and phase matching. It is a promising mid- and far-infrared NLO crystal material. This fact was contrary to the experimental results of powder multiplication tests because the experimentally characterized Ga_2S_3 cubic lattice is theoretically unable to achieve phase matching. Shuiquan Deng's group at the Fujian Institute of Research on the Structure, Chinese Academy of Sciences, developed three low-symmetry Ga_2S_3 structural models by designing a high-throughput calculation software, which denied the X-ray conclusion about the Ga_2S_3 cubic phase. The experimental measurements of multiplication factors and phase matching were explained by these structural models (Fig. 22).

Investigating the relationship between structure and NLO performance is crucial to the design of high-performance NLO crystal materials. Guocong Guo's group at the Fujian Institute of Research on the Structure, Chinese Academy of Sciences,

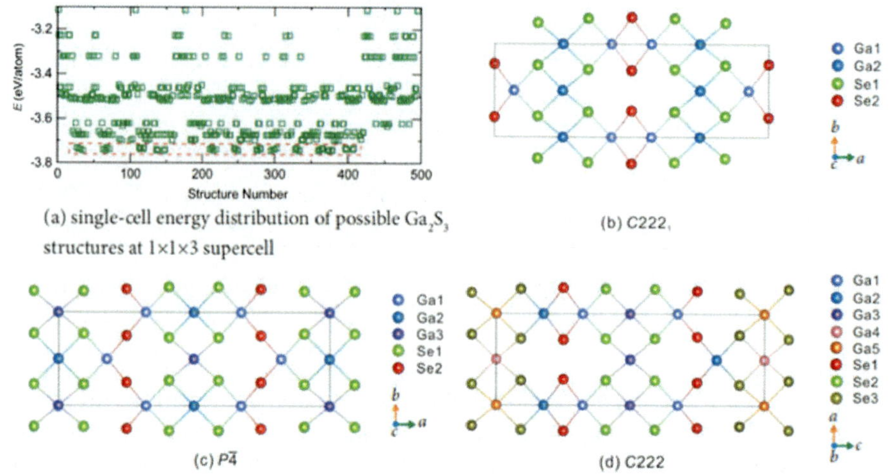

Fig. 22 High-throughput calculations to predict Ga_2S_3 crystal structures

studied the structure-effect relationships of NLO crystal materials from the perspective of the experimental electronic structure of "functional motifs". First proposed in 2001, the term "functional motif" refers to the microstructural unit that plays a key role in material functions. The orderly assembly of functional motifs can lead to high-performance materials. NLO functional motifs are structural units with large microscopic NLO polarizabilities in the material structure, contributing mainly to the macroscopic NLO effect of crystals.

To overcome the structural design bottleneck of high NLO coefficients and high laser damage thresholds, Guocong Guo's group at the Fujian Institute of Research on the Structure, Chinese Academy of Sciences, adopted a dual "functional motif" structural design concept under the guidance of the functional-motif idea, i.e., the idea of assembling "laser-induced damage threshold (LIDT) functional motifs" (polycationic groups composed of elements with great electronegativity differences to increase the band gap and thus the laser-induced damage threshold) and the "NLO active functional motifs" (introducing covalent-based structural units to increase the NLO coefficient) into a center-less structure at the molecular level to successfully synthesize a series of high-performance infrared NLO such as $[A_3X][Ga_3PS_8]$ (A = K, Rb; X = Cl, Br) (Fig. 23).

These compounds have high NLO coefficients (4 – 9 times that of $AgGaS_2$) and high powder laser damage thresholds (31 – 39 times that of $AgGaS_2$), thus breaking the bottleneck that it is difficult to combine a high NLO coefficient and a high laser damage threshold. For example, $Li[Cs_2LiCl][Ga_3S_6]$, $[ABa_2Cl][Ga_4S_8]$ (A = Rb, Cs), etc.

The NLO response is the second-order polarization of the material's electrons under the laser's action. The electronic structure of the NLO material changes in response to the external laser field. The electronic structure of NLO functional motifs in the same material has the largest response to the laser field. Under the funding of

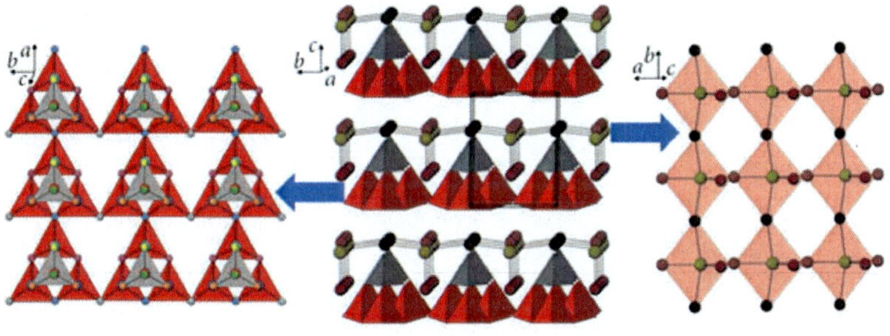

Fig. 23 Designing the infrared NLO crystal material$[A_3X]$ $[Ga_3PS_8]$ (A = K, Rb; X = Cl, Br) using the dual-functional-motif assembly strategy

the National Major Research Instrument Development Project, the research group of Guocong Guo at the Fujian Institute of Research on the Structure, Chinese Academy of Sciences, developed a new instrument for experimental electronic structure testing, which is fine-tuned to obtain the experimental electronic structure of the material using the X-ray diffraction method. They proposed a method of experimental determination of NLO functional motifs, i.e., to test the electronic structure (electron density distribution and wave function) of NLO crystals without laser (initial state) and with laser (functional state) in-situ, compare the topological characteristics, and classify the microstructure units with large changes in electronic structure as NLO functional motifs. The functional state herein refers to the state in which the electron cloud of the NLO crystal is polarized or distorted under the action of a laser (the laser wavelength is much smaller than the band gap of the crystal), showing the NLO function that is different from the excited state generated by the electron excitation jump in fluorescent materials and photoelectric conversion materials.

Guocong Guo's group at the Fujian Institute of Research on the Structure, Chinese Academy of Sciences, studied the in-situ electronic structure of the well-known NLO crystal LiB_3O_5 (LBO) under no light and 360 nm and 1,064 nm laser irradiation and found that the topological atomic charge, atomic volume and dipole moment of the $[B_3O_5]^-$ group changed significantly under laser light. Also, the electrons on the O atom were transferred to the B atom. Furthermore, the electronic structure of the BO_3 triangular unit responded more to the external fields than the BO_4 tetrahedron, while the electron cloud around Li changes negligibly. Therefore, the B-O group $[B_3O_5]^-$ can be determined as the NLO functional motif of LBO. This was the first study to experimentally test the electronic structure of the NLO material of LBO in the initial and functional states with high precision, study the electronic structure changes, reveal the NLO functional group, and make a leap from the atomic to the electronic level in the experimental study of the material structure, thereby providing a new way to study the functional group and the structure-effect relationship of the NLO material. This was the first international in-situ experimental electronic structure work in nonlinear optics, and was praised by the reviewers as a milestone in the field of NLO (Fig. 24).

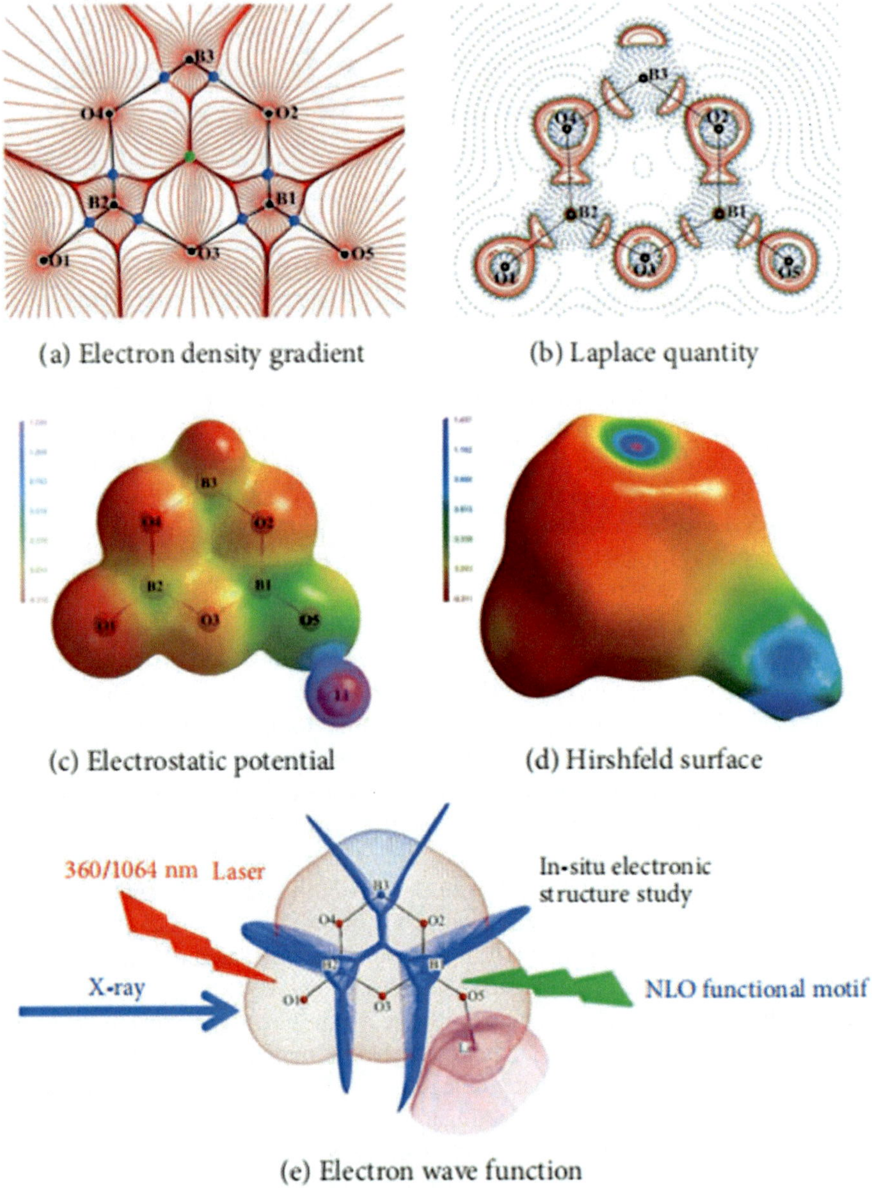

Fig. 24 In-situ experimental electronic structure of the "made-in-China" NLO crystal LBO

3.2 Deep-Ultraviolet Nonlinear Optical New Crystals

Deep-ultraviolet (Deep-UV) ($\lambda < 200$ nm) NLO crystals are the core components for obtaining all-solid-state deep-UV lasers, and so far, only $KBe_2BO_3F_2$ (KBBF) crystals have achieved direct hexafrequency deep-UV laser ($\lambda = 177.3$ nm) output from Nd:YAG; however, the severe lamellar habit has restricted the commercial production and practical application of KBBF. For decades, the design and synthesis of the next generation of deep-UV NLO crystals has been a high-priority of research. Using the KBBF structure as the design template, Ning Ye's research group at the Fujian Institute of Research on the Structure, Chinese Academy of Sciences, synthesized two deep-UV NLO crystals, $NH_4Be_2BO_3F_2$ (ABBF) and Be_2BO_3F (γ-BBF), which can effectively overcome the lamellar habit by strengthening the interlayer connection through N–H…F hydrogen bonding and Be-F ionic bonding while maintaining the structural and performance advantages of KBBF (Fig. 25). The UV cutoff, birefringence, and frequency multiplication effects of ABBF and γ-BBF are very close to or better than those of KBBF, and the shortest phase-matching wavelengths of class I can reach 173.9 nm and 146 nm, respectively. They are promising deep-UV NLO crystals, showing excellent deep-UV optical output potentials. The research results have been selected for the Innovation Achievement Exhibition of the Chinese Academy of Sciences (Fig. 26).

The research group of Shilie Pan at the Xinjiang Technical Institute of Physics and Chemistry, Chinese Academy of Sciences, pioneered the strategy of introducing $BO_{4-x}F_x$ ($x = 1, 2, 3$, [BOF]) functional groups into the borate framework to design deep-UV NLO crystals. Studies showed that the introduction of F^-ions increased the anisotropy of the [BOF] group, which could increase the birefringence of the material and avoid the laminar growth habit while obtaining a large band gap. Based on this, the group successfully designed and synthesized a series of fluoroborates AB_4O_6F (A = NH_4, Na, Rb, Cs, K/Cs, Rb/Cs) and $MB_5O_7F_3$ (M = Mg, Ca, Sr). These materials broke the limits of the traditional short-wave NLO crystal materials (short UV endpoint-large multiplication response-moderate double refractive index). They were expected to achieve deep-UV laser multiplication output below 200 nm. The $NH_4B_4O_6F$(ABF) crystal was successfully designed and synthesized with a very short UV cutoff (156 nm), a large multiplication factor (3 KDP), and a moderate birefringence to meet the deep-UV phase matching (the shortest matching wavelength of 158 nm was calculated) (Fig. 27).

Meanwhile, the crystal structure of ABF is more compact, and the interlayer force is significantly enhanced compared with KBBF, thus eliminating the laminar growth habit and obtaining centimeter-sized crystals. Furthermore, its raw material does not contain highly toxic beryllium elements, and its multiplication factor is 2.5 times higher than that of KBBF, enabling it to be used in deep-UV laser light sources to obtain higher conversion efficiency. Since then, the group has discovered nearly 50 new fluoroborates. Ning Ye's group discovered the new compound $M_2B_{10}O_{14}F_6$ (M = Ca, Sr) at the Fujian Institute of Research on the Structure, Chinese Academy of Sciences. All the discovered fluoroborate crystals with nonlinear activity exhibit

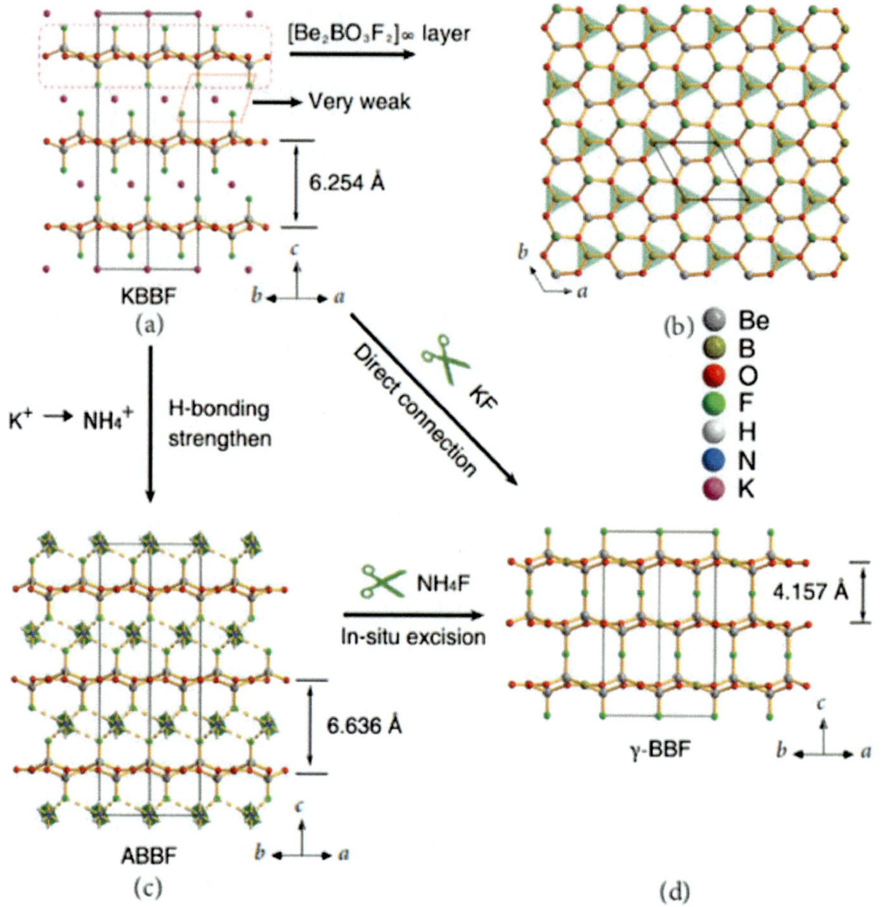

Fig. 25 Comparison of the crystal structures of **a** KBBF, **c** ABBF and **d** γ-BBF; **b** Functional layer of $[Be_2BO_3F_2]_\infty$

excellent NLO properties, achieving a balance between "wide band gap and large multiplication and suitable birefringence". This compound series has a more compact crystal structure with significantly enhanced interlayer forces compared with the KBBF crystal structure, thus reducing the laminar growth habit and making it possible to obtain larger crystals (Fig. 28). The above research results have been published in *Journal of the American Chemical Society* (2 articles), *Angewandte Chemie International Edition* (4 articles), and other international journals, featured twice in the *C&EN* journal, and selected as "Top 10 Optical Advances in China 2017".

Meanwhile, $K_3Ba_3Li_2Al_4B_6O_{20}F$ (KBLABF), a new beryllium-free short-wave UV NLO crystal, was designed and synthesized by Junhua Luo's group at the Fujian Institute of Research on the Structure, Chinese Academy of Sciences, using the well-known NLO crystal $Sr_2Be_2B_2O_7$ as the structural template and replacing the toxic Be^{2+} with Li^+ and Al^{3+} of similar coordination ability. The KBLABF lamellar units

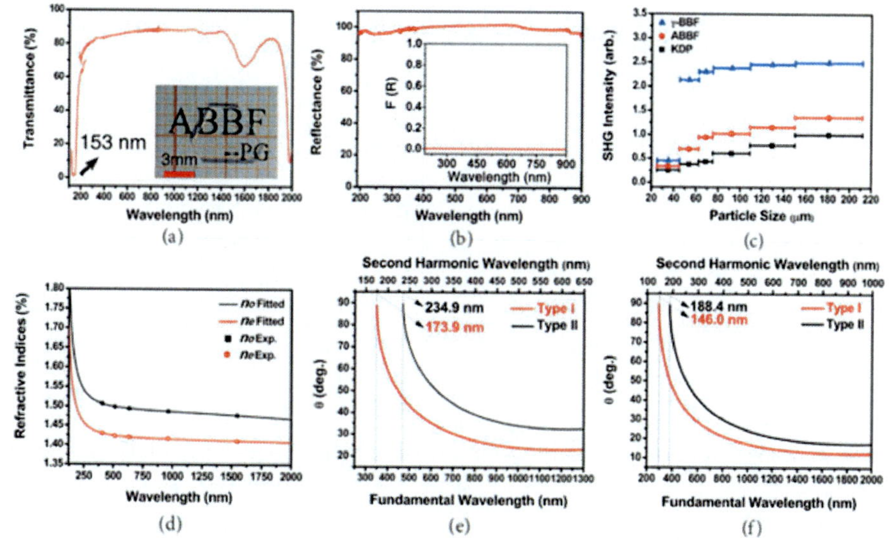

Fig. 26 **a** Transmission pattern of ABBF; **b** UV diffuse reflectance pattern of γ-BBF; **c** comparison of powder multiplication effect of ABBF and γ-BBF with the standard crystal KDP; **d** measured refractive index and refractive index fitting curve of ABBF; **e** phase-matching curve of ABBF; **f** phase-matching curve of γ-BBF

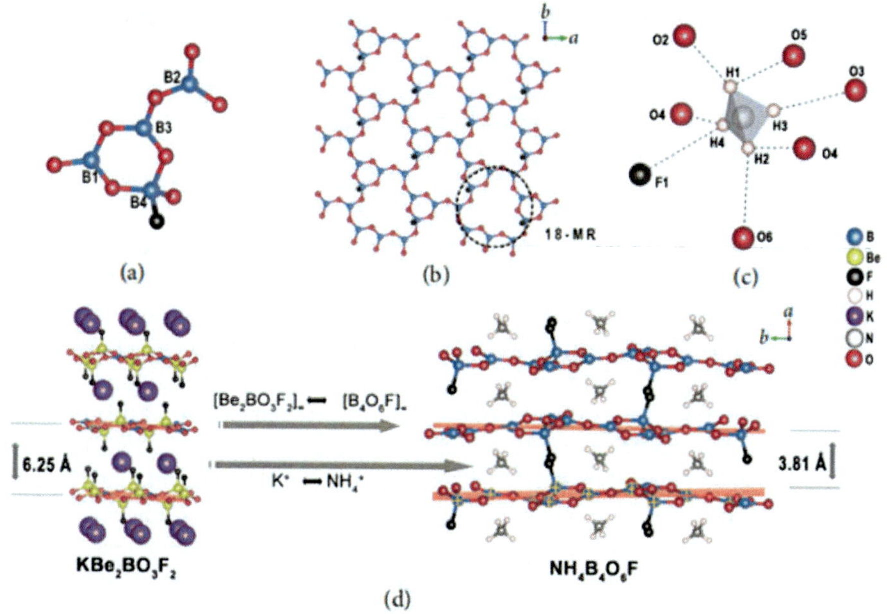

Fig. 27 **a** $(B_4O_8F)^{5-}$ microscopic motif in boron fluoride $NH_4B_4O_6F$ (ABF); **b** 2D $[B_4O_6F]_\infty$ layer; **c** NH^{4+} coordination environment; **d** Structural comparison of KBBF and ABF

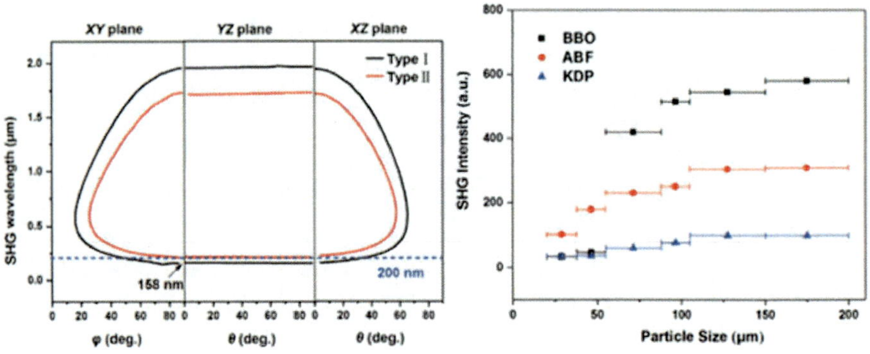

(a) (b)

Fig. 28 **a** Phase-matching wavelength of boron fluoride $NH_4B_4O_6F$ (ABF); **b** Comparison of ABF with standard crystals KDP and BBO for the effect of powder preparation

are firmly linked by Ba–O bonds, thus overcoming the lamellar growth habit and growing crystals up to 10 mm thick in the *c*-direction. In addition, the refractive index, laser damage resistance threshold, hardness, deliquescence resistance, and thermal properties of KBLABF were investigated. The crystal was found to have potential applications in multiplying the frequency output of solar-blind UV lasers. The results have been granted a Chinese invention patent and applied for a PCT international patent. Professor Poeppelmeier, former chairman of the Inorganic and Solid-State Chemistry Section of the American Chemical Society, commented that this work provided new opportunities for designing high-performance beryllium-free deep-UV NLO materials.

Traditionally, deep-UV NLO crystal materials have focused on typical functional motifs of π-conjugated systems, such as the BO_3 element of the off-domain π-conjugation. Junhua Luo's research group at the Fujian Institute of Research on the Structure, Chinese Academy of Sciences, proposed the concept of developing non-π-conjugated deep-UV NLO crystal materials and designed and synthesized a series of new non-π-conjugated deep-UV NLO crystal materials. Based on the self-polymerization strategy of tetrahedral motifs with a high multiplication factor and a short UV absorption edge, the group developed a new non-π-conjugated deep-UV NLO crystal material $RbBa_2(PO_3)_5$. Theoretical calculations of the optical properties of related materials were performed in collaboration with Zheshuai Lin from the Institute of Physical and Chemical Technology, Chinese Academy of Sciences, and it was found that the microscopic NLO coefficients of the corresponding phosphorus-oxygen structural elements tended to increase with the higher degree of polymerization of PO_4 tetrahedra. This gave rise to a new concept for designing and synthesizing new NLO crystal materials with large NLO effects. A new non-π-conjugated deep-UV NLO crystal $Ba_5P_6O_{20}$ with a significant blue-shifted UV absorption edge was successfully obtained based on the molecular tailoring design of the flexible structural motifs, and a blue-shifted UV absorption edge mechanism in contrast to that of KBBF was discovered. A new non-π-conjugated deep-UV NLO crystal $RbNaMgP_2O_7$ with an enhanced multiplication effect induced by thermal phase transition was synthesized, providing a physical methods to enhance the multiplication effect of NLO crystal materials compared with the conventional idea of chemically synthesizing NLO materials. Furthermore, the research group discovered two sulfate non-π-conjugated deep-UV NLO crystals, $(NH_4)_2NaLi_2(SO_4)_2$ and $(NH_4)_2Na_3Li_9(SO_4)_7$, and revealed that the anomalous multiplication effect between the two materials is attributed to the different orientations of the non-bonded oxygen atom 2p orbitals of the SO_4 motif in their structures, which provides an important reference for the structural design of these materials. In addition, inspired by the idea of "Yin/Yang Harmony" in traditional Chinese culture, the research group proposed the idea of introducing both the most electronegative F^- and the most electronegative Cs^+ to assist in the construction of non-centrosymmetric structures and successfully designed and synthesized the first fluorophosphosilicate non-π-conjugated deep-UV NLO crystal $CsSiP_2O_7F$. Theoretical calculations were performed to prove that $SiP_2O_{10}F$ is a novel NLO functional motif in its structure. This work has opened up a new system of fluorophosphosilicate NLO crystal materials.

Simultaneously, ultra-high peak power lasers can provide unprecedented extreme physical experimental conditions. Currently, international ultra-short pulse devices with a power output greater than 10 PW mainly use the OPCPA (optical parametric chirped-pulse amplification) technology. Large-size LiB_3O_5 (LBO) crystals are an irreplaceable core material for developing ultra-high peak power lasers of 10 – 100 PW class power using OPCPA technology. Recently, large-size LBO crystal devices have gradually become the "stranglehold" material for the whole device, restricting the development of ultra-high peak power lasers in China. The group of Xifa Long at the Fujian Institute of Research on the Structure, studied the evolution of components, flow, and temperature fields in the growth process of large-size LBO crystals by combining theories and process simulations, established the defect formation law, as well as the optimal control strategy and method for the large-diameter LBO crystal process. During the growth process of large-size LBO crystals, the research group gradually explored the optimization of the flux combination to solve the problems of high melt viscosity, thick growth boundary layer, and slow solute transport caused by the formation of the three-dimensional network structure of boron-oxygen bonding chain (O-B-O). They proposed and realized the process of growing large-size LBO crystals in selective orientation near-matching direction and solved the difficulties of growing large-size LBO crystals in the near-matching direction. The group optimized the growth system of large-size LBO crystals by studying the evolution of components, flow, and temperature fields in the crystal growth process, eliminating the heterogeneous crystals and wrapping them in the crystal growth process, and completed the development of the annealing process for large-size LBO crystals, and successfully grew 4,520 g of LBO single crystals (Fig. 29) with a crystal size of 240 mm × 180 mm × 80 mm. After processing, the LBO crystal device with a size of 110 mm × 120 mm × 26 mm was obtained and tested.

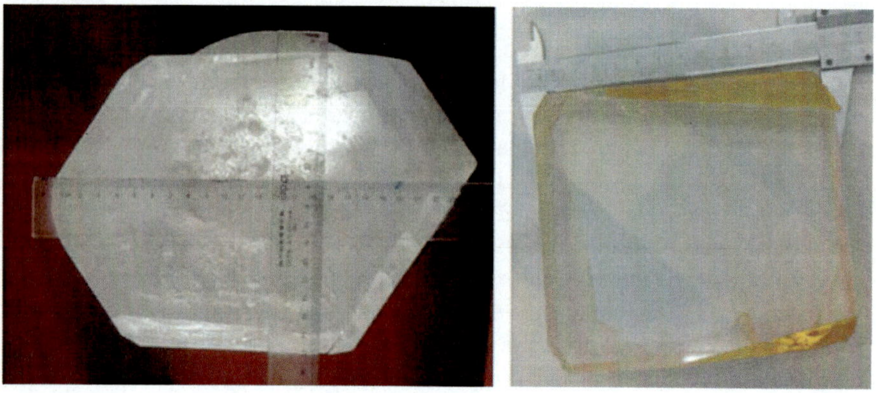

Fig. 29 Large-size LBO single crystal

3.3 Infrared Nonlinear Optical Crystal

Tunable infrared lasers have various military and civil applications. Currently, using the infrared NLO crystal parametric oscillation method is an important way to obtain infrared laser output, which has the advantages of compact laser structure, all-solid-state, high-power, narrow linewidth laser output, etc. The infrared NLO crystal is one of the most important components. However, the performance of the traditional chalcopyrite-based infrared NLO crystals $ZnGeP_2$, $AgGaS_2$, and $AgGaSe_2$ discovered in the 1970s abroad, cannot meet the development needs of modern infrared laser technology. Based on the energy-band modulation strategy, Ning Ye's group and Yao Jiyong's group in Fujian Institute of Research on the Structure, Chinese Academy of Sciences, and Institute of Physical and Chemical Technology (IPT) discovered two new high-performance infrared NLO crystals, $BaGa_4S_7$ and $BaGa_4Se_7$, respectively, by introducing the heaviest alkaline earth metal Ba into sulfur compounds. Currently, $BaGa_4S_7$ crystals with Φ 15 × 40 mm^3 have been stably grown by Ning Ye's group (Fig. 30), and $BaGa_4Se_7$ crystals with Φ 40 × 150 mm^3 have been grown by Jiyong Yao's group. The crystal has been used in a series of laser experiments in the United States, Germany, and Russia. The results showed that $BaGa_4S_7$ and $BaGa_4Se_7$ are a new class of mid- and far-infrared NLO crystals with excellent overall performance and important application prospects. In particular, the laser damage thresholds of these two materials are the highest among mid-infrared crystals, which solves the application bottleneck of low laser damage thresholds of infrared NLO crystals, marking this research work at the forefront of the world.

The paradoxical relationship between the wide band gap and the large multiplication effect of IR NLO crystals makes it a great challenge to explore the balanced performance of IR NLO crystals. The atoms in diamond structure compounds are all in a tetracoordinate environment, which is conducive to generate large band gaps. At the same time, these tetrahedral motifs are spatially arranged in cubic or dense hexagonal stacks with the same orientation, which is conducive to the generation of strong

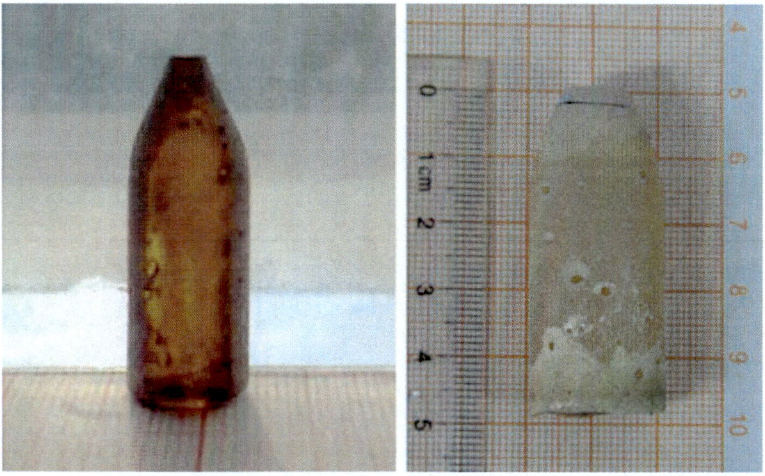

Fig. 30 $BaGa_4S_7$ large-size single crystal

multiplication effects. Therefore, the group of Zheshuai Lin at the IPT, Chinese Academy of Sciences focused on the study of sulfur-like diamond compounds, mapped a complete family of diamond-like structures, and performed large-scale calculations of their properties. The study showed that the commercial AgGaS2 crystals are still the best performing conventional diamond-like IR nonlinear crystals. Still, some defective diamond-like structures exhibit significantly enhanced multiplication effects while retaining a band gap close to that of AgGaS2. Subsequently, Jiyong Yao's group at the IPT, Chinese Academy of Sciences, experimentally discovered defective Hg2GeSe4 crystals with a multiplication effect twice that of AgGaSe2 crystals and proved their theoretical prediction. A new diamond-like infrared NLO crystal, Li4HgGe2S7, with a multiplication effect 1.5 times that of AgGaS2 crystals and a damage threshold 3.5 times that of AgGaS2, was also reported by the group of Shili Pan at the Xinjiang IPT, Chinese Academy of Sciences. Using the classical sphalerite and sillimanite structures as templates, the group of Ning Ye at the Fujian Institute of Research on the Structure, Chinese Academy of Sciences, successfully obtained four examples of NLO crystals of iodosubstituted phosphorus compounds, namely M3PnI3 (M = Zn, Cd; Pn = P, As), by introducing the strongly electronegative heavy halogen I into phosphorus compounds using a heterovalent anion substitution strategy (Fig. 31). They have a defective diamond structure with a consistent arrangement of [MPnI3] mixed anion groups within the structure, which feature a strong multiplication effect ($2.7 - 5.1 \times AgGaS_2$), a large band gap (2.38 – 2.85 eV) and a wide infrared transmission range, striking a balance among band gap, multiplication effect, and infrared transmission range. These studies suggested that diamond-like structures are a potential direction for exploring new types of excellent infrared NLO crystals.

In the application development of mid- and far-infrared NLO crystal materials, Guocong Guo's group at the Fujian Institute of Research on the Structure, Chinese

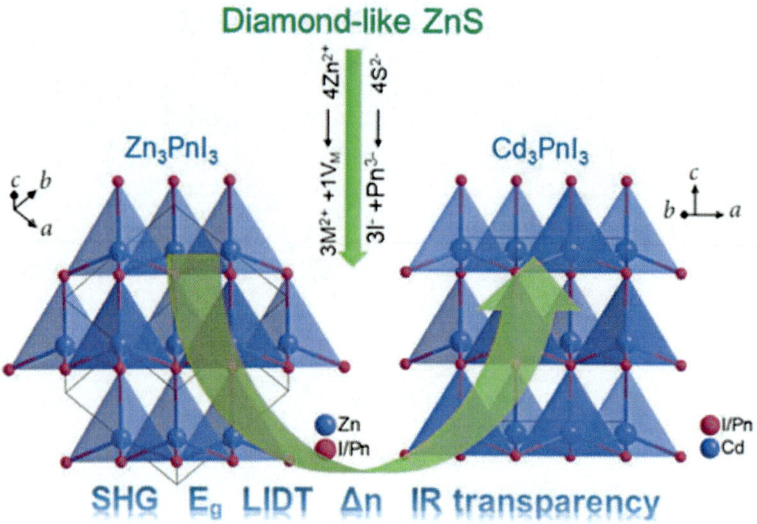

Fig. 31 Crystal structure of M_3PnI_3 (M = Zn, Cd; Pn = P, As)

Academy of Sciences, discovered monoclinic phase (β-)Ga_2S_3 NLO crystal materials with a high laser damage threshold (30 $AgGaS_2$), a SHG response comparable to $AgGaS_2$ (0.83 $AgGaS_2$), a wide transmission range (0.4 – 12.5 μm), and 1.064 μm pumping that can achieve phase matching within 3.15 – 12 μm. It is one of the very few NLO crystals that can achieve phase matching in both 3 – 5 μm and 8 – 12 μm IR windows, with a specific heat capacity higher than $AgGaS_2$, a birefringence $\Delta n = 0.0239$, slightly higher than $AgGaS_2$, and homogeneous melting, showing excellent overall performance and great application potentials. The results have been granted two Chinese patents and two US patents. Guocong Guo's group at the Fujian Institute of Research on the Structure, Chinese Academy of Sciences, leveraged an innovative crystal growth technique for Ga_2S_3 crystals in high-temperature hexagonal phase, medium-temperature monoclinic phase, and low-temperature cubic phase, respectively, and made a breakthrough in growing the Ga_2S_3 crystals in medium-temperature monoclinic phase. They grew crystals larger than 30 mm and manufactured crystal devices of 5 mm × 5 mm × 5.08 mm (Fig. 32).

3.4 *Laser Crystals and Transparent Ceramics*

The 1.55 μm band laser, which is safe for human eyes, and located in a good atmospheric transmission window and the detection sensitivity region of Ge and InGaAs detectors, can be widely used in LIDAR laser ranging and 3D imaging. Er^{3+}/Yb^{3+} double-doped phosphate glass is the only commercially available 1.55 mm band laser material. However, the glass's low thermal conductivity and laser damage threshold

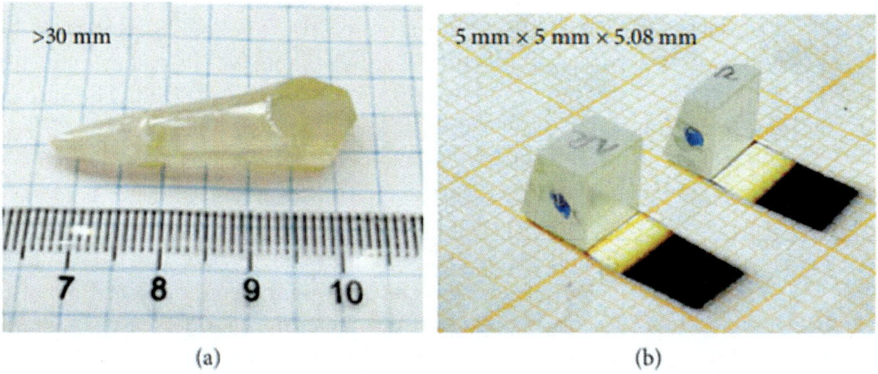

Fig. 32 **a** Ga2S3 crystal; **b** Device

make it difficult to achieve high average power laser output. The 1.55 mm band laser devices developed based on this material cannot fully meet the application needs of the relevant devices. Based on the study of the structure-effect relationship between the crystal structure and spectrum and laser performance, Yidong Huang's research group at Fujian Institute of Research on the Structure, Chinese Academy of Sciences, developed a new Er^{3+}/Yb^{3+} double-doped $RAl_3(BO_3)_4$ (R = Y, Gd, Lu) series of laser crystals by modulating multiple facets of laser crystals with multiple light-emitting centers. They also grew single crystals with large sizes (>45 mm × 45 mm × 30 mm) and high optical quality by improving the flux system and optimizing the growth process parameters. Using these crystals as the working materials, the group achieved high-efficiency (>35%) and high-power (>2.0 W) continuous laser operations in the 1.55 mm band and developed high-performance pulsed laser devices with internationally leading performance. The human-eye safe 1.55 mm band miniature solid-state laser based on these crystals is expected to provide an excellent detection light source for autonomous driving LIDARs and laser rangefinders. The research work not only protects China from being contained by other countries in self-driving LIDARs and laser rangefinders but also enables us to counter them.

In the area of composite functional crystals, Jiyang Wang's group at Shandong University focused on the combination of laser and frequency-multiplying functions. They successfully coupled the laser with two core functional motifs of NLO crystals (anionic group and multi-luminous center polyhedra), designed a new laser self-multiplication crystal Yb/Nd:Y(Gd) COB, completed the growth of large-size laser self-multiplication crystals and design of devices, and achieved the highest power laser self-multiplying green light output in the world. The group solved the long-standing problem that the output efficiency of self-multiplying crystals was lower than that of laser and frequency-multiplying discrete devices. The crystal is highly recognized and evaluated in several projects, and achieved industrialization, which has a significant share in the international small power green laser market, driving the application of downstream products to a market size of 1 billion yuan. Meanwhile, it is the first in the world to create a yellow self-multiplying laser device, achieving a 10W-class laser output and filling the gap in practical yellow laser devices. It is expected to be used for laser illumination and indication, etc. Its patent rights have been transferred, and the initial industrialization has been realized. The research results won the second prize of the National Invention Award in 2012.

In addition, to prepare large-size, high-quality laser transparent ceramics, Yongge Cao's group at the Fujian Institute of Research on the Structure, Chinese Academy of Science used the non-equilibrium preparation theory and surface activation to prepare monodisperse, spherical, high-purity YAG ceramic nanopowders with high sintering activity, which significantly reduced the sintering temperature, avoided abnormal grain growth, achieved nearly "porosity-free" sintering, and decreased scattering losses of ceramics. By exploring the combination of suitable sintering aids and the theory and technology of grain boundary engineering and crystal field modulation, combined with the thermodynamic and kinetic control of the sintering process of ceramic billets, the effective regulation of grain boundaries and microstructures was achieved, and thin, straight and pure grain boundaries and uniform grains with very few microporosity were obtained, which further reduced the inherent scattering loss of ceramics (reaching 0.2% of that of single crystals). Based on this, the research group successfully produced large-size high-quality transparent ceramic with a diameter of 350 mm. In 2012, they achieved 3300 W continuous laser output from a single Nd:YAG ceramic, marking China the third country in the world to achieve a laser output of more than one kilowatt from a single ceramic piece. In 2013, a 5,000 W heat capacity laser output was further achieved. We were the first in the world to report the successful production and laser output of Yb:YAG laser transparent ceramics with a sandwich composite waveguide structure using the flow casting method, which further achieved the highest laser efficiency of 70% for Nd:YAG ceramic.

In conclusion, we have systematically explored the theory of structure-effect relationships of NLO crystals, NLO crystals in the deep-UV and mid-infrared, and special-wavelength laser crystals and achieved the goal of structural design and controlled synthesis of optoelectronic functional substances. The research on UV NLO crystals keeps leading the world, and some research on infrared crystals has entered the international frontier; breakthroughs in crystalline materials have solved several "stranglehold" technological problems in several laser application fields.

4 New Electron-Doped Iron-Selenium-Based Superconductors

New high-temperature superconductors are not only the basis for solving the major cutting-edge scientific problem of high-temperature superconductivity mechanisms but also have important applications in information, energy, and medical fields. However, the discovery of new high-temperature superconductors is very difficult.

It took 22 years, from the discovery of copper-based high-temperature superconductors in 1986 to the discovery of iron-arsenic-based ones high-temperature superconductors in 2008, mainly due to the lack of new functional motifs to construct superconductors. In February 2008, Hosono's research group in Japan first achieved superconductivity at 26 K in F-doped LaOFeAs, immediately attracted widespread attention in the condensed matter physics community. Later, Zhongxian Zhao's group and Xianhui Chen's group in China soon raised the critical temperature of FeAs-based superconductors to above 40 K, breaking the MacMillan limit and proving them as non-conventional superconductors. Currently, the transition temperature of Fe-As-based superconductors has reached a maximum of 56 K, making them the second most important family of superconductors after Cu-based superconductors in terms of the transition temperature. The existence of other types of high-temperature superconducting materials besides Cu-based and Fe-As-based superconductors has become an important frontier in the field of superconducting materials research.

4.1 First Discovery of Electron-Doped Iron-Selenium-Based Superconductors

Since the discovery of iron-arsenic high-temperature superconductors, a series of iron-based superconductors with different structures have been discovered. Typical systems include ReFeAsO (Re = rare-earth elements) (1111 system), AFe_2As_2 (A = K, Sr, Ba, etc.) (122 system), LiFeAs (111 system), etc. Except for FeAs-based superconductors, β-FeSe has structural elements similar to FeAs layers. Still, its crystal structure consists only of FeSe layers composed of co-sided tetrahedra stacked along the *c*-axis, without the charge reservoir layers used to provide carriers in other FeAs-based superconductors. The tetragonal FeSe superconductor transition temperature is about 8 K at atmospheric pressure.

Based on the idea of structural motif regulation, the research group of Xiaolong Chen at the Institute of Physics, Chinese Academy of Sciences, successfully discovered a new FeSe superconductor, $K_xFe_2Se_2$, with a critical superconductivity temperature of 30 K. The average structure of the tetragonal FeSe superconductor has been identified to be body-centered. The space group is *I4/mmm* (Fig. 33). Compared with FeAs-based superconductors, $K_xFe_{2-y}Se_2$ is significantly different in structure and properties, mainly in (i) the electronic structure, where only electron-type Fermi surfaces exist in the Brillouin zone, and hole-type Fermi surfaces are far away from the Fermi energy level, indicating that its superconducting transition mechanism is different from other Fe-based superconductors; (ii) the structure of the new superconductor has K and Fe vacancies, and the Fe vacancies undergo a transition from disordered to ordered at 576 K, leading to multiple superstructures; (iii) an antiferromagnetic transformation accompanies the disordered-ordered transformation with a magnetic moment of Fe up to 3.3 μB/Fe, which greatly exceeds the magnetic moment of Fe in Fe-As-based superconductors.

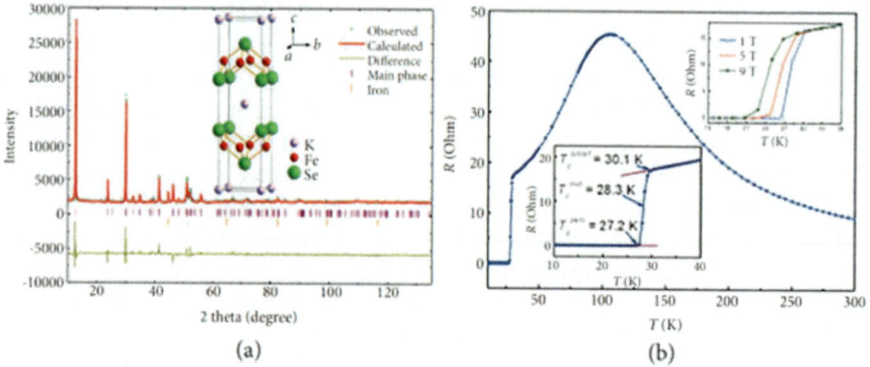

Fig. 33 **a** X-ray diffraction pattern and Rietveld refined spectrum of $K_{0.8}Fe_2Se_2$, the inset shows the crystal structure of $K_{0.8}Fe_2Se_2$; **b** Variation of resistance of $K_{0.8}Fe_2Se_2$ with temperature, the inset is a magnified view of the low-temperature region

Several research groups at home and abroad have demonstrated that the Fermi surface configuration of $K_xFe_2Se_2$ is completely different from that of Fe-As-based high-temperature superconductors, challenging the mainstream belief that it is the Fermi surface nesting that induces superconductivity and triggers a new wave of superconductivity research. This research has opened up a new direction in superconductivity research and was ranked as 1st and 7th most popular research in physics by Thomson Reuters in "Research Fronts 2013" and "Research Fronts 2014." This new type of iron-based high-temperature superconductor, which was discovered for the first time by Chinese scientists, was recognized by domestic and foreign scholars that it can be comparable with the Fe-As-based high-temperature superconductors. This work has had a lasting impact, and several important results have been obtained based on it by researchers both in China and abroad.

4.2 *First Discovery of Molecular Intercalation Series of Iron-Selenium-Based High-Temperature Superconductors*

The isostructural superconductors $Cs_{0.8}(FeSe_{0.98})_2$, $Rb_xFe_ySe_2$, $Tl_{0.58}Rb_{0.42}Fe_{1.72}Se_2$ and $(Tl,K)Fe_xSe_2$ with similar properties were synthesized successively after the discovery of $K_xFe_{2-y}Se_2$. Numerous experimental methods have proved the prevalence of phase separation in this series of superconductors, i.e., the coexistence of the antiferromagnetic primary phase $A_2Fe_4Se_5$ (245 phase) with the superconducting phase due to the ordering of iron vacancies. 245 phase is insulating and does not represent superconductivity. Currently, superconducting phases are generally considered to be streaked phases embedded in the 245 matrix phase precipitated from $A_xFe_{2-y}Se_2$, and their phase content is usually low, about 10% – 20%. These superconducting phases are isomorphic to $BaFe_2As_2$ and can be expressed by the chemical formula $A_xFe_2Se_2$ ($x = 0.3 - 1.0$), but the reported x values vary widely. In addition, several research groups in China and abroad have tried to obtain $K_{0.8}Fe_2Se_2$ isomorphic superconductors by substituting alkali or other metals other than K, Rb, and Cs, but in vain. It is imperative to obtain their pure superconducting phases or true single crystals to clarify the special electronic structure of these superconductors and study their intrinsic properties.

By inserting alkali metals Li, Na, alkaline earth metals Ca, Sr, Ba, and rare-earth elements Eu and Yb into FeSe layers at room temperature, the research group of Xiaolong Chen at the Institute of Physics, Chinese Academy of Sciences, innovated the low-temperature liquid ammonia method, and discovered and prepared a series of $A_x(NH_3)_yFe_2Se_2$ (A = Li, Na, Ca, Sr, Ba, Eu, Yb) high-temperature superconductors that cannot be obtained using conventional high-temperature methods. At atmospheric pressure, the highest superconducting critical temperature is 46 K, which exceeds the McMillan limit and is the highest value for FeS-based superconductor blocks. The discovery of $A_x(NH_3)_yFe_2Se_2$ series superconductors provides a new starting point for exploring superconducting materials and studying the mechanism of high-temperature superconductivity.

4.3 *True Composition and Structure of Superconducting Phases in the K-Fe-Se System*

After the first report of $K_xFe_2Se_2$ with a superconducting transition temperature of 30 K, the prevalence of phase separation in this series of superconductors has seriously affected the study of the intrinsic properties of this superconducting system. Various attempts by other international research groups to obtain pure superconducting phases in the K-Fe-Se system using Bridgman's or other high-temperature methods have failed. In addition to the 30 K superconducting transition, the 44 K superconducting transition is often observed in some samples of the K-Fe-Se system, further complicating the problem. Since the main phase of the so-called $A_xFe_{2-y}Se_2$ "single crystal" samples is the 245 phase with a very low content of superconducting phases, the previous physical reports are questionable. Clarifying the true composition and structure of the superconducting phase in the K-Fe-Se system is an important research topic.

The group of Xiaolong Chen at the Institute of Physics, Chinese Academy of Sciences, successfully used the liquid ammonia method to precisely regulate the doping amount of alkali metals and suppress the phase separation, and identified the true composition and structure of the superconducting phases in the K-Fe-Se system for the first time to obtain the superconducting phases of K-Fe-Se series superconductors. It was found that there were at least two $ThCr_2Si_2$ superconducting phases with complete FeSe layers in the potassium intercalated FeSe compound (Fig. 34). The group further found that the 44 K superconducting phase is transformed

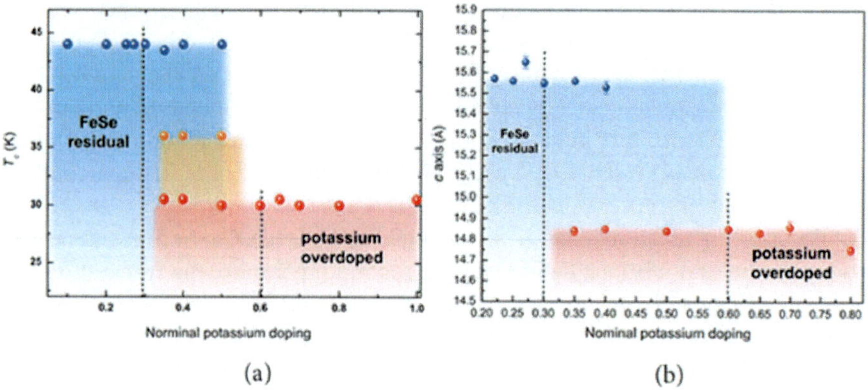

Fig. 34 **a** The trend of T_c of $K_xFe_2Se_2(NH_3)_y$ with the nominal content of potassium; the variation of T_c with the amount of potassium doping is discrete and different from that of Cu- or FeSe-based superconductor, which shows a "dome" shape; **b** The trend of the lattice constant c of $K_xFe_2Se_2(NH_3)_y$ with the nominal content of potassium, with two discrete c's; when $0.3 < x < 0.6$, two different lattice constants c coexist

to the 30 K superconducting phase after increased potassium doping. This leads to the conclusion that the 30 K phase is overdoped. More importantly, nitrogen removal experiments showed that nitrogen only affects the lattice constant but has almost no effect on the superconducting transition. Therefore, the above conclusions also fully apply to $K_{0.3}Fe_2Se_2$ and $K_{0.6}Fe_2Se_2$ superconductors. This work clarifies the composition and structure of the superconducting phase in the potassium-iron-selenium system and laid the foundation for further studies.

4.4 *First Anderson Localization in the Single-Crystal Electron System*

More than half a century ago, Anderson proposed that disorder could cause the localization of electrons and spins. Anderson localization has had a wide and profound impact on several concepts and phenomena in physics, such as the quantum Hall effect, quantum critical points, random matrix theory, and electron interactions in disordered metals. Although Anderson localization was originally proposed for electrons, it is a wave localization phenomenon. It was first realized in systems such as photons, phonons, cold atoms, mechanical waves, matter waves, and low-dimensional electrons. Much of the early work focused on the metal–insulator (MIT) transition in heavily doped semiconductors for the three-dimensional electronic system. Still, the MIT transition was mainly associated with the transition from discrete energy levels of impurities to energy bands, i.e., the emergence of simplex semiconductors, which was not the ideal system for studying Anderson localization. Recently, Wuttig's research group in Germany found for the first time a disorder-only

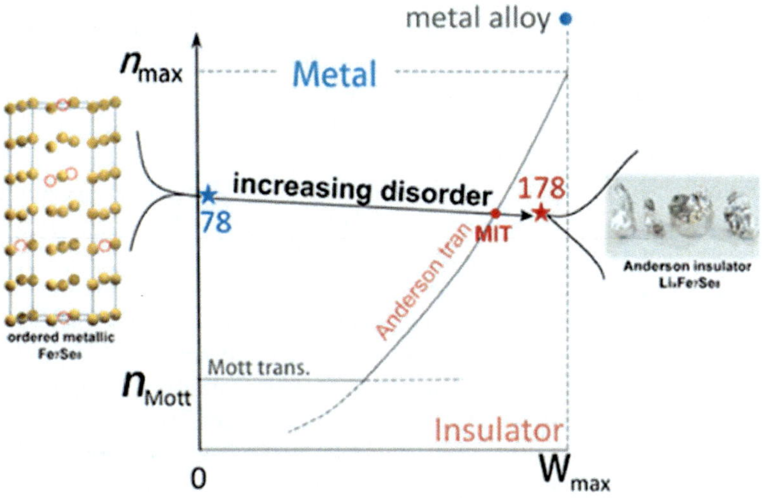

Fig. 35 Phase diagram of carrier concentration versus disorder, with the horizontal and vertical coordinates of disorder and carrier concentration, respectively

MIT transition in $GeSb_2Te_4$ polycrystals, where the localization of electrons was due to the disorderly distribution of vacancies in the Ge/Sb sublattice. However, due to the small grain size (10 ~ 20 nm), it was inconvenient to investigate the effect of lattice disorder on electron localization. To study the effect, it is an important research topic to realize the Anderson localization of the electron system in a three-dimensional single crystal.

The group of Xiaolong Chen at the Institute of Physics, Chinese Academy of Sciences, successfully obtained $Li_xFe_7Se_8$ in the centimeter-scale single-crystal form by electron doping the metallic parent phase Fe_7Se_8 and simultaneously inducing the disordered occupation of iron in the system and observed the Anderson localization of electrons. This work further gave a more generalized phase diagram of carrier concentration versus disorder (Fig. 35), which helped discover more Anderson insulators. This was the first report of Anderson localization of the electron system in a single bulk crystal, the results of which provide a new platform for studying disorder and Massachusetts Institute of Technlogy transition, deepening the understanding of the electronic behavior in disordered materials.

4.5 Regulation of Superconductivity in Thin-Layered Iron-Selenium (FeSe) Single Crystals

Chemical doping is a useful technique for introducing carriers into solid materials. In both Cu- and Fe-based superconductors, chemical doping can induce high-temperature superconductivity where antiferromagnetic ordering is suppressed in

Cu-based superconductors, and spin density wave ordering is suppressed in Fe-based superconductors. However, elemental substitution changes the carrier concentration to a very limited extent and introduces much disorder in the system. Therefore, many intrinsic phenomena have not been observed. Recently, applying FETs to modulate carrier concentration in 2D systems has become an effective way to control material properties. The use of electrostatic doping to regulate the carrier concentration can regulate the doping effect of novel phases on a large scale, which is generally difficult to achieve using conventional material synthesis methods.

Xianhui Chen's group at the University of Science and Technology of China (USTC) successfully regulated the superconducting phase of FeSe flake samples by using the ionic liquid gate voltage technique. The group achieved superconductivity up to 48 K at the upper critical temperature in FeSe samples with only $T_c = 10$ K by mobilizing the gate voltage to the electron-doped side. This is the first time that superconductivity above 40 K was achieved in FeSe flakes without an interface or applied pressure, indicating that high-temperature superconductivity of 48 K can be induced in FeSe with only a simple electron doping process. Interestingly, the data suggested that the evolution from low to high T_c phase is a Lifshitz phase transition that occurs abruptly at a particular carrier concentration (Fig. 36). These results are useful for building a unified picture of the heavy high-temperature superconductivity of FeSe-based superconductors and beneficial for exploring higher T_c in similar materials in the future.

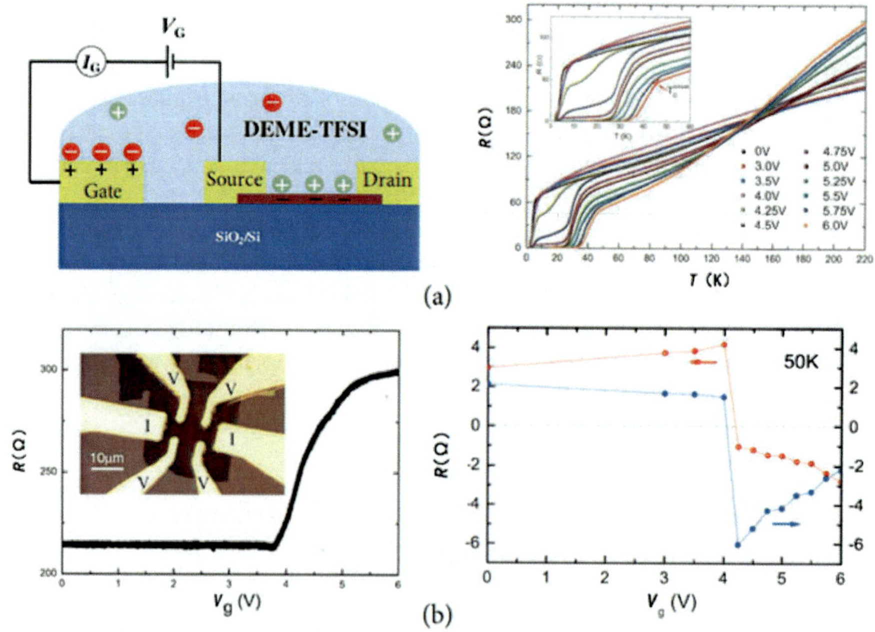

Fig. 36 Evolution of superconducting phase in FeSe regulation by ionic liquid gate voltage technique

The discovery of the novel FeSe-based high-temperature superconducting materials described above has led to rapid development in the research field of high-temperature superconductivity and has played a key role in the further understanding of non-traditional high-temperature superconductivity mechanisms. The discovery of $K_xFe_2Se_2$ was featured in *American Physical Society* and *Materials Research Society*, respectively. It was highlighted by *Nature Materials* in its editorial commemorating the centennial of superconductivity discovery, stating, "What's now for the new century? More superconductors will be discovered. Recently, superconductivity with T_c around 30 K in the new family of Fe-based materials $A_xFe_2Se_2$ (A = K, Cs) was reported. The observation is generating excitement, especially as new results point toward different electronic and magnetic properties from all other Fe-based superconductors". Thomson Reuters ranked "alkali-doped iron-selenide superconductors" among the top 10 most active frontiers in physics in *Research Frontiers 2013*, recognizing China's strengths in this frontier area. In addition, Prof. Loidl of Augsburg University, Germany commented that the discovery of $A_x(NH_3)_yFe_2Se_2$ "presents a completely new route for iron-selenide superconductors". Dr. Conder of the Paul Scherrer Institute in Switzerland called the discovery of $A_x(NH_3)_yFe_2Se_2$ "the first report of an intercalated iron-selenium superconductor of alkali and alkaline earth metals from a liquid ammonia solution". To date, more than 350 research groups in 41 countries and regions have conducted follow-up studies of electron-doped iron-selenium-based superconductors. The discovery of $K_xFe_2Se_2$ has been cited 944 times. In 2020, to mark the 50th anniversary of the journal, the editors of *Physical Review B* selected the 50 most influential papers from more than 190,000 published articles; the discovery of $K_xFe_2Se_2$ was the only research completed entirely by Chinese scientists.

5 Other Novel Crystalline Materials

5.1 Metal–Organic Framework Isoporous Materials

China has made a number of significant research advances in the field of porous materials, particularly MOF materials, including the separation and purification of light olefins, CO_2 capture, and catalytic conversion of small molecules, among others, all of which have had a significant academic impact on the international arena, with some of the results leading international research.

We have developed new concepts and methods for separating and purifying small molecule olefins by adsorption. Small molecule olefins, such as ethylene, propylene, and 1,3-butadiene, are the most important raw materials for chemical products, synthetic plastics, and rubber in the world. They usually come from the gas from petroleum cracking, which is a mixture of olefins and alkanes with very similar properties, requiring a separation and purification process with high-energy consumption. It is possible to greatly reduce the energy consumption via absorption separation using porous materials. However, the adsorption using traditional porous materials, which preferentially adsorbs more polar olefins, does not simplify the separation process and makes it difficult to obtain high-purity olefins. A new MOF material, MAF-49, was synthesized by Jie peng Zhang's group, who first proposed a strategy to adsorb ethane using ultra-microporous hydrophilic porous materials selectively. The preferential adsorption of ethylene/ethane mixture through a fixed adsorption bed composed of MAF-49 at room temperature and pressure resulted in the preferential capture of ethane due to the non-classical hydrogen bonding in the confined space of the ultra-microporous pores, and the high-purity ethylene could be obtained directly (Fig. 37). In one adsorption process, 1 L of MAF-49 can produce 56 L of ethylene with > 99.95% purity.

Based on the ethylene/ethane separation, Jiepeng Zhang's group further proposed a strategy to control the conformation and adsorption enthalpy of flexible guest molecules by using quasi-isolated pores and non-classical hydrogen bonding in restricted space to reverse the adsorption selectivity of C_4 hydrocarbons. They validated the effectiveness of the new strategy through experimental and theoretical simulations of a series of representative porous coordination polymers and selected C_4 hydrocarbon material MAF-23 for optimal separation. After passing a mixture of C_4 hydrocarbons through a fixed-bed adsorption device filled with MAF-23 at ambient temperature and pressure, butadiene, usually the most readily adsorbed one, inversely becomes the least adsorbed component and is thus the first to flow out. The purity met the requirements of the subsequent polymerization reaction (>99.5%), effectively eliminating the current risk of self-polymerization in the purification of butadiene (Fig. 38).

Industrial propylene is usually made by cracking propane, and removing leftover propane often necessitates a large capital investment and a large amount of energy. Weigang Lu and Dan Li's group suggested a new orthogonal array dynamic sieving

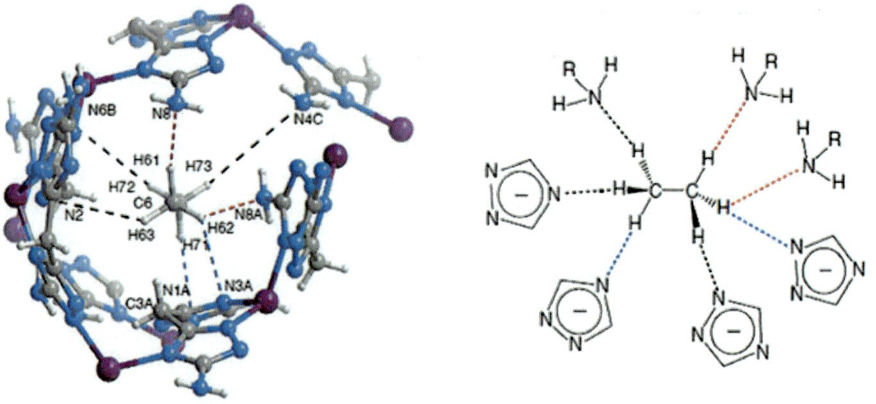

Fig. 37 Non-classical hydrogen bonding of ethane molecules with functional groups in the ultra-microporous hydrophilic MAF-49 material

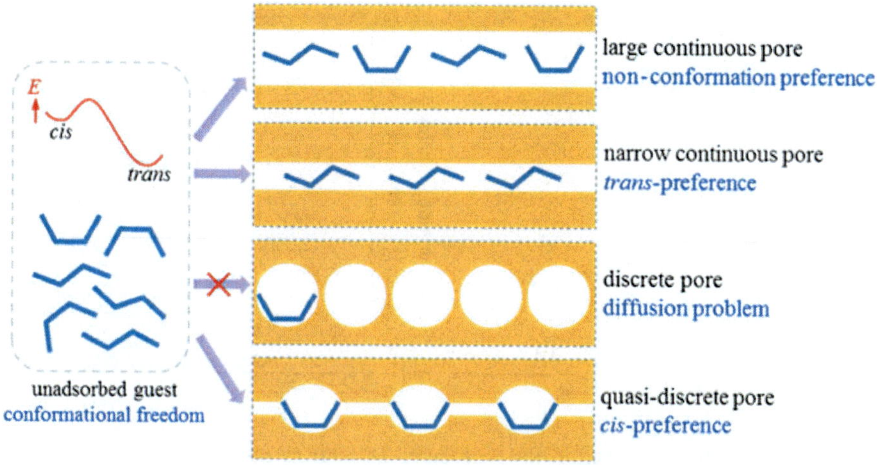

Fig. 38 Inversion principle of adsorption selectivity by controlling the guest configuration using gated pore channels

mechanism that successfully solved the fundamental shortcomings of classic molecular sieves' delayed adsorption kinetics and limited adsorption capacity (Fig. 39). Based on the new sieving mechanism, a breakthrough in propylene/propane separation was achieved by synthesizing MOF materials that can rapidly separate propylene/propane (1:1) mixtures.

53.5 L of polymerically pure (99.5%) propylene was obtained from 1 kg of MOF materials. The team also developed a new sieving mechanism and dynamic process induced by the molecular interaction between propylene/propane molecules and materials through in-situ single-crystal diffraction and molecular simulation. The research results provide a green solution for the efficient separation of propylene/

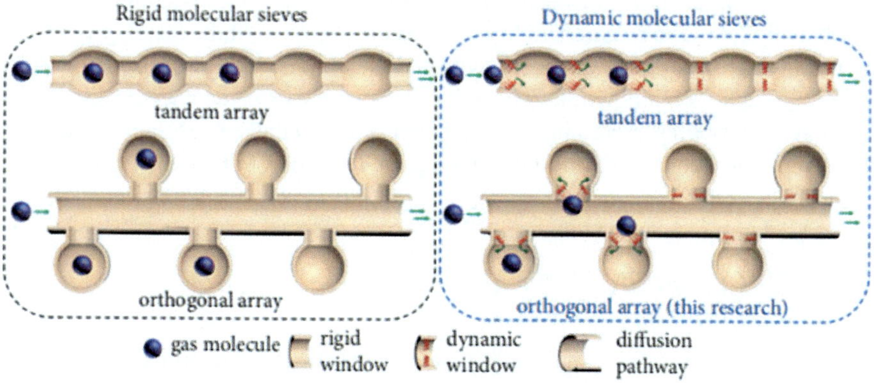

Fig. 39 Schematic illustration of tandem and orthogonal array sieving mechanism

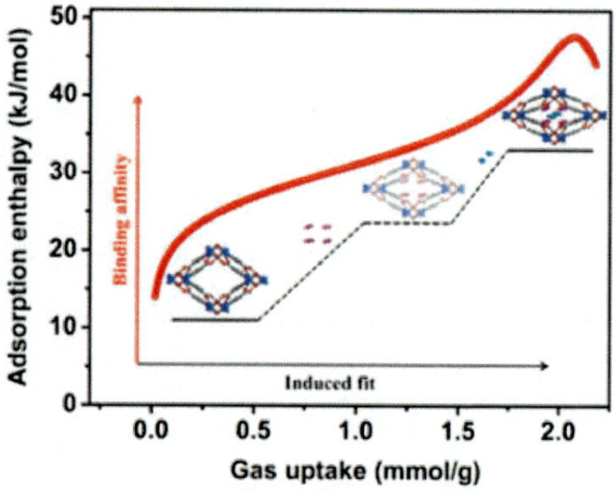

Fig. 40 Schematic representation of the flexible MOF material induced-fit effect

propane and other important gas molecules as well as new ideas for the design of next-generation novel MOF sieving materials.

Weigang Lu and Dan Li's research team introduced the induced-fit effect of enzymes into the MOF materials. The enthalpy of acetylene adsorption was observed to increase paradoxically with acetylene adsorption in a flexible MOF material with one-dimensional rhombic pores, and the induced fit to acetylene was obtained by opening the metal sites. The acetylene/carbon dioxide penetration experiments revealed that the acetylene/carbon dioxide mixture (50:50) passed through the activated framework material solid filled column with a penetration time of 75 min for carbon dioxide and 150 min for acetylene, demonstrating its excellent separation performance for the two materials (Fig. 40).

Cellular metabolism is frequently accompanied by the passage of ions and tiny molecules across the membrane in living organisms (e.g., protons, calcium ions, sodium ions, potassium ions, water molecules, etc.). Nature has inspired scientists to create artificial molecules or units that imitate this critical biological function. The team of Xiaoping Zhou and Dan Li successfully constructed a rhombic dodecahedral metal–organic cage with a window size very similar to the dynamic diameter of carbon dioxide. The researchers finely tuned the pore size of the metal–organic cage by changing the strategy of substituents and metal ions. It allows carbon dioxide to enter through its window into the cavity of the metal–organic cage or escape from the cavity through pressure induction, which achieves the encapsulation and release of carbon dioxide, successfully mimicking the function of the alveoli (Fig. 41).

5.2 2D Crystalline Materials

Firstly, the group of Yi Xie at the USTC was the first to develop a universal method for the exfoliation of layered hybrid intermediates (Fig. 42) and to prepare a series of 2D ultrathin crystalline materials with non-layered crystallographic characteristics. The

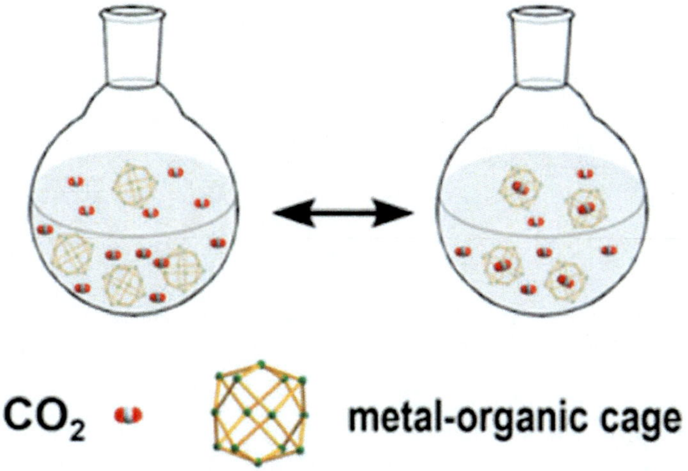

Fig. 41 Pressure-induced CO_2 coating/release

group also proposed the first ligand-assisted bottom-up method to prepare 2D ultrathin crystalline materials for non-lamellar compounds with no or weak anisotropy. The group developed substituted solid-solution exfoliation and lithium-insertion-delithiation methods for chemically bonded layered compounds to produce 2D ultrathin crystalline materials. The group has developed a series of 2D ultrathin crystalline materials using the direct liquid-phase exfoliation approach for layer-like compounds with van der Waals forces between the layers.

Secondly, the group was the first to employ synchrotron X-ray absorption fine structure spectroscopy to resolve the atomic structure and different coordination environments of the surface atoms of 2D ultrathin crystalline materials (Fig. 43). The research results were selected as "Major Achievements of Major Science and Technology Infrastructure of the Chinese Academy of Sciences". The group introduced the positron annihilation technique to characterize the defect structure of 2D ultrathin crystalline materials. It accurately characterized their defect types and contents, based on which an accurate database of structural parameters and structural models

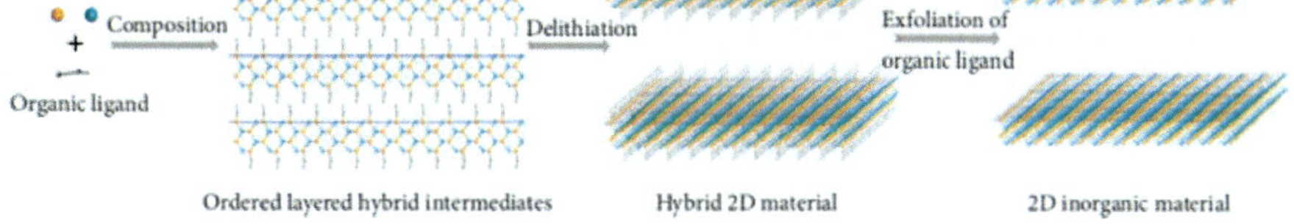

Fig. 42 Schematic diagram of the universal preparation strategy for 2D inorganic materials with specific atomic layer thickness

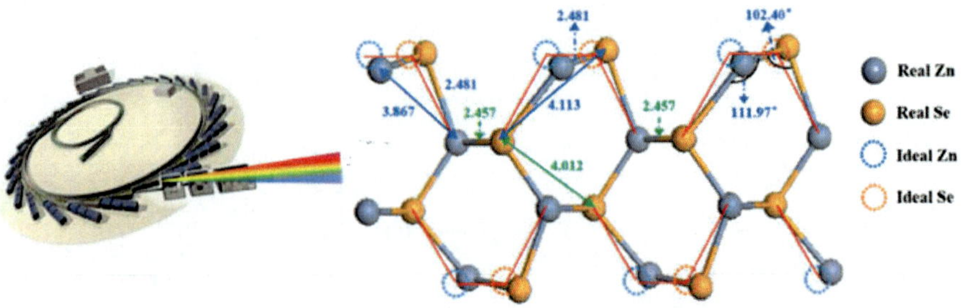

Fig. 43 Fine structure of 2D inorganic materials resolved at the atomic scale: synchrotron X-ray absorption fine structure spectroscopy of the 2D atomic structure of four-atomic-layer-thick ZnSe

was established. The technique rapidly became one of the novel tools for studying the microstructure of low-dimensional solid materials.

Furthermore, the group has discovered that 2D ultrathin crystalline materials exhibit a peculiar electronic state with a substantially larger density of states near the Fermi plane than the bulk material using first-principles calculations based on density generalization theory. All properties based on electron leap or electron transport are expected to be improved by using this type of material. To improve their photoelectric catalytic properties, the group further modulates the electronic structure of 2D inorganic ultrathin crystalline materials by creating defects such as vacancies, doping, surface modification, and structural hybridization. For example, to address the key scientific challenges in the efficient and targeted conversion of CO_2 to carbon-based fuels, the research group pioneered the use of 2D ultrathin crystalline materials with high activity, high density, and high homogeneous surface sites as an ideal model system to regulate and optimize the performance of CO_2 photo/electrocatalytic conversion. The research work on 2D ultrathin cobalt-based catalysts elucidated a new mechanism for efficient activation of carbon dioxide by special electronic states in ultrathin structures, and realized the reduction of carbon dioxide in the low-energy barrier pathway (Fig. 44). The research on "a new cobalt-based electrocatalyst for efficient and clean conversion of carbon dioxide into liquid fuels" was selected as one of the 2016 top 10 advances in Chinese Science. Further, to improve the selectivity of CO_2 reduction products, a new mechanism of the 2D ultrathin bimetallic sulfide was designed and prepared, and a new mechanism of the two active sites to change the CO_2 reduction pathway was elucidated to achieve nearly 100% selectivity of the reduction products, which provided a new idea for the construction of highly selective CO_2 reduction catalysts.

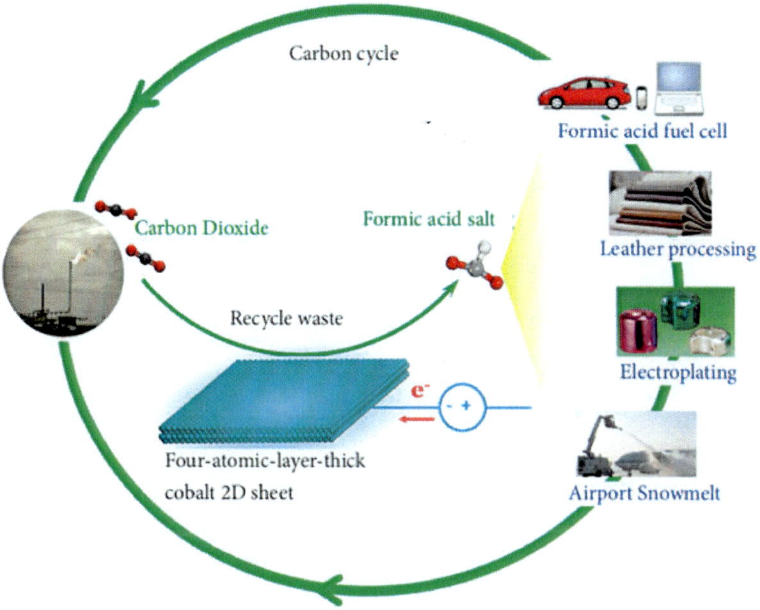

Fig. 44 Green and economical carbon cycle by electrocatalytic CO_2 reduction

Outlook

Maochun Hong, Yicheng Wu, Chunhua Yan, Yuliang Li, Jiyang Wang, Xianhui Chen, Jingjun Xu, Yongjun Chen, Rong Chen, Baosheng Huang, Xuefeng Fu, Junlin Yang, Shouzhu Zhang, Kexin Chen, Qidong Wang, Jie He, Hongcheng Lu, and Guohong Zou

China has gradually established a relatively complete research system for crystalline novel materials during the implementation of this Plan. China has assembled a research team with advanced academic ideas and innovative ability at the international frontier in the field of crystalline materials such as laser crystals and NLO crystals, molecular ferroelectrics, molecular magnets, MOF materials, superconducting materials, and thermoelectric materials. An interdisciplinary system with coordinated and complementary cooperation has been established, and academic leaders who are forward-thinking, competent, capable, and enterprising in this field

M. Hong (✉)
Fujian Institute of Research on the Structure of Matter, Chinese Academy of Sciences, Fuzhou, China
e-mail: hmc@fjirsm.ac.cn

Y. Wu
Tianjin University of Technology, Tianjin, China

C. Yan
Peking University, Beijing, China

Y. Li
Institute of Chemistry, Chinese Academy of Sciences, Beijing, China

J. Wang
Shandong University, Jinan, China

X. Chen
University of Science and Technology of China, Hefei, China

J. Xu
Nankai University, Tianjin, China

Y. Chen · R. Chen · B. Huang · X. Fu · J. Yang · S. Zhang · K. Chen · Q. Wang · J. He · H. Lu · G. Zou
National Natural Science Foundation of China, Beijing, China

© Zhejiang University Press 2023

M. Hong (ed.), *Structural Design and Controllable Preparation of the Function-Directed Crystalline Materials*, Reports of China's Basic Research,
https://doi.org/10.1007/978-981-99-3768-4_4

have emerged. We have made ongoing efforts in the past 10 years to lead in several important frontiers of new material, providing practical results for China to lead basic research and the historic opportunity of "overtaking" the development under the guidance and integrated organization of NSFC and the steering group of experts.

Nevertheless, due to the wide scope of advanced materials research, new materials are still emerging. Our basic and independent innovation ability is still weak, and the slow transformation of research results remain a prominent problem in this field. In addition, the research groups of new crystalline materials are widely distributed, and although there are many prominent individual advantages, the overall sustainable development advantages are lacking.

At present, there still exist following shortcomings.

(i) There are many sources of innovation, but they are not coordinated, and the development momentum still needs to be stimulated. During the Plan, there have been leading achievements in some points, but they have not been connected organically. In particular, there are many high-level articles and a few core patents. Even in the field of materials in which we are leading, some key materials and devices used in practice are often restricted. Therefore, while researching materials with new properties, we must focus on the breakthrough of key and highly demanded "stranglehold" crystalline materials.

(ii) Insufficient understanding and attention to the whole chain of new material design, exploration, preparation, characterization, and application. As the relationship between the material motif, structure, function, and service performance gradually deepens, the basic research is more in-depth, but insufficient attention is paid to the basic research on the materials and devices highly demanded by the nation. In the integrated research, insufficient effort is being made to focus on the laser crystals and NLO crystals and their preparation technologies with significant application backgrounds to make breakthroughs. It is necessary to develop a theoretical system with Chinese characteristics and enter the "free" stage from "inevitable" development. It is imperative to promote basic research on material preparation further, especially to strengthen the understanding of the development of the whole chain, and to strengthen the overall planning of applied basic research and its industrialization, to promote the long-term sustainable high-level development and overall improvement.

(iii) The development of basic theoretical research, new methods, and new technologies in material preparation science is weak. Currently, materials research tends to only focus on the materials themselves; there is insufficient attention paid to the common scientific research on materials preparation. There is little research on new methods, new technologies, and equipment for materials preparation, and insufficient investment, affecting the momentum of materials development. The new technology of "controllable preparation" needs to be further developed, and the development of key preparation technologies of key materials should be given special attention.

(iv) Insufficient innovation, long cycle time, and low efficiency is shown in new material development patterns. Material development has long used discrete disciplines, single simulation experience, or traditional trial-and-error models, with long development cycle and low efficiency. The domestic computational materials science team is scattered. There is a lack of computational software with independent intellectual property rights, and the computational materials science and materials engineering applications is not closely integrated. In materials research and design, emphasis is placed on calculating electronic and related properties, whereas less attention is paid to artificial intelligence and quantum computing in materials design and controlled preparation.

(v) The integration of multidisciplinary intersection, basics, and application needs further improvement. This Plan attaches great importance to interdisciplinary research, especially the intersection of chemistry and materials, chemistry and physics, and materials and physics. In the early stage, scholars in chemistry synthesized a large number of new compounds with novel structures and excellent properties and developed effective and innovative preparation methods; scholars in physics discovered new physical phenomena and explored new laws; scholars in materials science developed promising materials systems and engineering technologies. However, on the one hand, due to limited budgets, the research in the form of incubation projects and key projects is, to a large extent, an internal cycle of research carried out by the project leaders in their organizations, so it is objectively difficult to conduct further interdisciplinary research related to function-oriented crystalline materials. On the other hand, since the cross-disciplinary research atmosphere is still being cultivated, and the attribution of cross-disciplinary research results and the scientific and technical evaluation of project applications are still under continuous improvement, there is a large room for collaboration among chemistry, materials, and physics, and the interdisciplinary research teams, which are hard to organize, need continuous incentives from the Plan management and funding. This requires support with high-intensity funding for integrated projects and a modest increase in funding for the Plan with a developmental mindset to guide and strengthen substantive out-of-cycle collaborative research in interdisciplinary areas, including collaboration and research with international counterparts. Furthermore, due to limited budgets, the major international collaborative projects and related shared platforms and databases planned in the implementation plan have not yet been implemented.

1 Strategic Needs

The accelerated application of artificial intelligence and quantum computing in the research and development of new cutting-edge materials and the extensive interdisciplinary research between materials science and other disciplines will further expand the research and application of new cutting-edge materials. Therefore, it is imperative to continuously support, develop and expand the overall advantages and, meanwhile, strengthen the application and transformation of key materials, which can solve the basic research problems in application and transformation.

The long research and development chain of new crystalline materials, includes material preparation and its mechanism, crystalline structure and property characterization, component processing, and device preparation, as well as the design and exploration of new crystalline states, fully reflecting the interdisciplinary integration of the materials science, condensed matter physics, solid-state chemistry, and mechanical engineering. Therefore, a single research institution or a scattered research team cannot meet the development needs of functional information materials in the new era.

At present, microelectronics and photonics are closely integrated and are moving toward the era of photonics. Optoelectronic and photonic materials will become the fastest growing and most promising information materials, and the development of advanced crystalline functional materials plays a key leading role in the future development of science and technology. The crystalline state used for optoelectronic components is developing toward large-size, high uniformity, and high lattice integrity, while the components themselves are developing toward thin-film, multi-function, chip, ultra-high integration, and low-energy consumption; microelectronics technology continues to reduce the size of devices and increase the chip area to improve the integration and information processing speed, from single-chip integration to system integration. Optoelectronics is developing toward nanostructure, non-uniform, nonlinear, and non-equilibrium states. Optical integration will be an important direction for the development of electronic technology in the next decade and beyond.

In recent years, high-power laser materials, optical communications, optical sensing, high-density storage, micro, and nano manufacturing support materials, passive electronic components, and other optoelectronic materials are developing rapidly and applied worldwide. The research and development of disruptive electronic materials and devices, mainly advanced crystalline materials, are attracting more and more attention. The emerging photonic integration technology is an effective way to solve this problem. The next-generation memory technology will evolve toward high speed, small size, low voltage, high density, low power consumption, low cost, and system integration, etc. New crystalline materials such as magnetic functional molecular crystalline materials, single-molecule devices, superconducting, ferroelectric, and thermoelectric materials developed in this Plan will be the basis for the expected development.

In the twenty-first century, the mature development of quantum theory and the advancement of various characterization techniques have enabled us to understand and study materials on the microscopic subatomic scale. The correlations between different scales and dimensions of materials and the rich derivative phenomena and synergistic phenomena they lead to can be observed and discovered. New research fields such as quantum information, quantum computing, and topological physics have been opened up. At the same time, with the continuous improvement of diverse preparation technologies, new materials with various functions and even unprecedented properties have been created, including completely artificial materials. The discovery of many new materials and their new properties and physical phenomena are integrated with the cutting-edge physical theories, driving further improvement of physical theories. Meanwhile, advanced physical theories guide new physical materials' design, preparation, and application, which are being used to develop a series of new materials and devices to meet the need of technological development and improve human life.

Crystalline materials, as the carriers of functional properties, have always played a dominant role among new materials and devices. Crystalline materials with stable and orderly crystal structures, show rich physical connotations and various macroscopic functional properties in optics, electricity, and magnetics due to its diversity and easy compound and regulation, and are widely used in information, energy, medical and defense fields, bringing revolutionary progress to the industrial structure and human life. Currently, mankind has entered the information society and the era of big data. The traditional microelectronics field has shifted from "microelectronics" to "nanoelectronics" and from "Moore's law era" to "post-Moore era." Microelectronic devices and integrated circuits must go from single device size miniaturization to functional integration. There is an urgent need for new materials, devices, and designs to break the bottleneck of traditional CMOS (complementary metal–oxide–semiconductor), such as resistive memory, new display, and ultra-wide band semiconductor materials and devices to adapt to higher speed and more intelligent needs. New flexible electronic devices (including wearable electronic devices, etc.) must be multifunctional, thin, flexible, and even intelligent. Flexible electronic materials and devices are widely used in information, energy, medical, and defense applications and will bring revolutionary progress to the industrial structure and human life. The development of flexible multifunctional-integrated electronic devices depends on developing new flexible, functional materials and cutting-edge manufacturing technologies. It is also conducive to meet the needs of the energy and information industries for next generation technologies.

In the field of laser crystals and NLO crystals, there is important demand for laser applications in extended wavelengths. Lasers of various wavelengths are needed, from deep-UV and infrared to quantum communication. Based on the research of functional motifs, we will further strengthen the synergy between various functional motifs and different functional motifs, develop new mechanisms, design new materials, and establish new technologies to obtain new functional crystals and their applications in mid- and far-infrared, communication wavelengths, and solar-blind area as needed. The NSFC approved this research project as a major research project in 2018. The chemical design and ferroelectric coupling of molecular ferroelectrics further describe in the language of chemistry the fundamental theories related to ferroelectric phase transitions, such as Landau's theory of phase transitions and Curie's Symmetry Principle, and Neumann principle. A targeted "quasi-spherical theory" strategy is proposed for the directional design of molecular ferroelectrics. Using structure and symmetry control, the structural design and controllable preparation of ferroelectric function-oriented crystalline materials can be realized and the "H/F substitution" can be enriched and improved. The "H/F substitution" is expected to realize the controlled introduction and regulation of chirality and the construction of molecular ferroelectrics with a single chirality, which will greatly enrich the application prospects of molecular ferroelectrics. At the same time, the "H/F substitution" can achieve a significant increase in T_c and P_s through proper molecular designs, which can lay a solid foundation for the preparation of high-performance molecule-based ferroelectrics. Based on the controlled synthesis of molecular ferroelectricity, the next-generation molecular piezoelectric materials that reach or even surpass the performance of inorganic molecule-based piezoelectric materials can be obtained by using symmetry design and polar axis control, making it a useful complement to inorganic piezoelectric materials. The NSFC approved this research as a Plan in 2019.

In the future, China will focus on the precise synthesis, multi-stable construction, and regulation of new molecular system magnetic quantum materials, synthesize high critical temperature molecular magnets, high blocking temperature SMMs and 4d/5d heavy transition system single-chain magnets, and explore the structure–property relationship and regulation of molecular magnetic materials. In addition, under the main research guidelines of constructing new magnetic structures and developing new materials, we will explore the applications of molecular magnetic materials in spin manipulation, quantum coherence, quantum entanglement, information storage, and molecular spintronics by combining physical and material science methods and structural simulations, and finally develop characterization and evaluation tools to construct molecular spintronic devices for practical applications.

In superconductivity, the United States and Japan are the traditional powerhouses in copper oxide and iron-based high-temperature superconductivity research. They led the discovery of copper-based and iron-arsenic-based superconductors, respectively. Although China is fast catching up and then competing in the superconducting temperature of Cu- and Fe-As-based high-temperature superconductors and is playing an original and leading role in the fields of electron-doped FeSe superconductors and FeSe/STO interface superconductivity, we still lack important original efforts at the Nobel Prize level. In addition, in the field of theoretical and critical experimental measurements of superconductivity mechanisms, China has made overall progress compared with the earlier stage but still lags behind the United States. It is necessary to maintain continuous investment in this direction to be the first to crack the fundamental problem of unconventional superconductivity mechanisms in physics.

In the field of superconductivity research, China is a leader in applying iron-based superconducting materials. However, there is a big gap between the United States and Japan in applying practical devices, i.e., complete sets of equipment. In this direction, Chinese scientists are at the forefront of the world in the theories of conventional high-voltage superconductors, but further breakthroughs are needed in the experimental research, expecting to lead the research in the field of room-temperature superconductivity. The practical application with large market demand is expected to see an increase in investment from domestic universities and institutes.

In the field of thermoelectricity, the research and development of new semiconductor thermoelectric materials in China have reached an advanced level globally. Still, the main focus is currently on the research and development of thermoelectric materials in middle and high-temperature regions. There is insufficient research and development of materials in the middle and low-temperature regions. In addition, there is a wide gap between China and the United States and other thermoelectric material powerhouses, especially in the research of high-efficiency thermoelectric devices. Therefore, these research areas still need continuous investment and research.

Quantum and optical computing will gradually replace microelectronics-based optical fiber communication and electronic computers. The research and application of new photonic materials should become the frontier and hot spot of optoelectronic information materials, especially the materials with excellent optical bistability and the corresponding new devices, which will be the basis and pioneer of new high-tech industry in the new era. Therefore, it is necessary to gradually develop materials and chips required for new quantum communication and photonic computing to solve the design and equipment problems for key materials and devices.

Advanced crystalline materials are the precursor and foundation for developing strategic emerging industries. With the help of modern high-tech industries such as modern communication, computer, information network technology, micromechanical intelligent systems, industrial automation, and home appliances, we should focus on making breakthroughs in the core technology of functional crystalline materials represented by microelectronics, optoelectronics, and functional materials for electronic components to ensure the effective supply of key materials. We will make efforts to improve our independent innovation capability, build a new material engineering science and technology innovation system with the support of science and technology and talents, support the large-scale engineering construction of materials needed by national major projects, focus on solving the problems of the large-scale production process, equipment technology and environmental protection facilities construction for such functional materials, and greatly improve the international competitiveness of our new material industry.

2 Conceptions and Suggestions

A new round of scientific and technological revolution is brewing, and we are in an era of integrated and converged knowledge when different disciplines and their internal branches depend more and more on one another. The functional crystalline materials, as a typical interdisciplinary subject, will have a more intense competition than ever before. To occupy a favorable position in the new round of revolution with the arrival of the era of convergence, we need to focus our efforts on joint research of several advanced crystalline materials that are closely related to the national strategic needs and the development of emerging industries, to make major breakthroughs, and support the continuous innovation and development of emerging industries such as information technology, high-end equipment and manufacturing, advanced medical care, and other major national projects in China.

From the perspective of stable disciplinary development, it is crucial to provide continuous support for the research and development of new high-performance materials and the breakthroughs in traditional materials that act as a bottleneck in national defense and economic construction. Through the keen judgment and ongoing efforts of chemists, material scientists, and physicists, China's chemistry and materials science research is at the forefront of the world, ranking second only to the United States in terms of the number of SCI-accepted papers. However, there is still a big gap compared with developed countries, regarding original and widely applied materials with independent intellectual property rights. In particular, key materials related to national defense and economic security are still restricted by other countries. Therefore, it is necessary to further support the research to maintain the current development trend, enhance the international influence of China's materials research, and support the sustainable development of China.

In recent years, the intersection and frontier of chemistry, materials, and physics are changing rapidly in the international arena. In response to this trend, this Plan has assembled an innovative research team under the "Structural Design and Controllable Preparation of Function-oriented Crystalline Materials" fund, with the particularly noteworthy participation of young scholars returning from Europe, the United States, and Japan. They have solid foundation in chemistry, materials science, and physics and are competent for high-level interdisciplinary research. After returning to China, most of them have conducted innovative research at the core and frontier of chemistry, materials, and physics and started to obtain internationally influential results. However, for various reasons, many excellent young scholars have yet to receive support from this Plan. Therefore, it is necessary to set up new incubation and focus programs to support outstanding young scholars to conduct research in this field to create a concentration effect of outstanding young scientific and technological talents on the major research platform of the Plan.

During the decades of implementing this Plan, China has made great progress in crystalline materials science and technology. With the increasing international competition in the field of crystalline materials, especially in the geopolitical situation where the United States has imposed a total embargo on China's "stranglehold" materials, China must further systematically carry out basic research on new functional crystalline materials based on the present world-renowned achievements in the development of key technologies related to national security. To this end, some suggestions are proposed as follows.

(i) To focus on basic research, enhance the innovation capability, maintain the characteristic advantages, and provide an inexhaustible source for developing functional crystalline materials in China. In particular, China should further strengthen the research on new crystalline functional materials and pay attention to the important role of artificial intelligence and quantum computing in the design of crystalline materials, functional-motif properties, controllable preparation, etc., to put an end to the passive state of China when it comes to "stranglehold" crystalline materials, make breakthroughs in the research of crystalline materials, create several new materials with independent intellectual property rights, and develop our research characteristics in the field of crystalline materials research, especially in the area of new functional crystalline states to achieve a new leap forward.

(ii) To further strengthen the intersection and integration of chemistry and materials science and simultaneously expand the intersection with other disciplines, especially condensed matter physics, information science, medicine, and biology, to develop new research and development fields of crystalline materials.

(iii) To strengthen top-level design, improve supporting policies, and accelerate the application of functional crystalline materials. From an overall perspective, the pattern of materials research driven by the application will not change. China should increase its investment in basic research on materials, make moderate advance arrangements, focus on promoting the development of the functional crystalline materials industry and the convergence of high-end manufacturing technologies, construct a core equipment technology platform, material preparation technology platform, and fine processing platform characterized by "automatic control + artificial intelligence (AC + AI)", develop a comprehensive capability throughout the "physical design-virtual verification-material processing-structural assembly-device fabrication-integrated packaging-characterization" flow, and shorten the application cycle of information function materials.

(iv) To make key breakthroughs and focus on overcoming several barriers to transforming materials with cutting-edge research results and significant application prospects. Relying on the advantageous position of China in some fields of functional crystalline materials and oriented by the world market in the background of the demand from the international optoelectronic industry, we should bring together our strengths and advantages, make key breakthroughs and concentrate our efforts on preparing some functional crystalline materials and devices that are of great significance to the development of national economy and the defense, to guarantee our economic development and national defense strategic security.

(v) To strengthen the research on basic science and technology of controllable devices, key preparation technologies, and related equipment, and to highlight the key points. To solve the problem of "stranglehold" key materials and devices and improve our ability to counter other counties, we need to make significant breakthroughs in the development of the whole chain and promote the transfer of scientific and technological achievements and the gathering of talents, thereby boosting the transformation of China from a large country to a strong country in materials.

Index

© Zhejiang University Press 2023
M. Hong (ed.), *Structural Design and Controllable Preparation of the Function-Directed Crystalline Materials*, Reports of China's Basic Research,
https://doi.org/10.1007/978-981-99-3768-4